White Rose Maths

White Rose Maths Edition

Power Maths

Year 4A
A Guide to Teaching for Mastery

Series Editor: Tony Staneff
Lead author: Josh Lury

Pearson

Contents

Introduction to the author team

Power Maths arises from the work of maths mastery experts who are committed to proving that, given the right mastery mindset and approach, **everyone can do maths**. Based on robust research and best practice from around the world, *Power Maths* was developed in partnership with a group of UK teachers to make sure that it not only meets our children's wide-ranging needs but also aligns with the National Curriculum in England.

Power Maths – White Rose Maths edition

This edition of *Power Maths* has been developed and updated by:

Tony Staneff, Series Editor and Author

Vice Principal at Trinity Academy, Halifax, Tony also leads a team of mastery experts who help schools across the UK to develop teaching for mastery via nationally recognised CPD courses, problem-solving and reasoning resources, schemes of work, assessment materials and other tools.

Josh Lury, Lead Author

Josh is a specialist maths teacher, author and maths consultant with a passion for innovative and effective maths education.

The first edition of *Power Maths* was developed by a team of experienced authors, including:

- **Tony Staneff and Josh Lury**

- **Trinity Academy Halifax** (Michael Gosling CEO, Emily Fox, Kate Henshall, Rebecca Holland, Stephanie Kirk, Stephen Monaghan and Rachel Webster)

- **David Board, Belle Cottingham, Jonathan East, Tim Handley, Derek Huby, Neil Jarrett, Stephen Monaghan, Beth Smith, Tim Weal, Paul Wrangles** – skilled maths teachers and mastery experts

- **Cherri Moseley** – a maths author, former teacher and professional development provider

- **Professors Liu Jian and Zhang Dan**, Series Consultants and authors, and their team of mastery expert authors: **Wei Huinv, Huang Lihua, Zhu Dejiang, Zhu Yuhong, Hou Huiying, Yin Lili, Zhang Jing, Zhou Da and Liu Qimeng**

  Used by over 20 million children, Professor Liu Jian's textbook programme is one of the most popular in China. He and his author team are highly experienced in intelligent practice and in embedding key maths concepts using a C-P-A approach.

- **A group of 15 teachers and maths co-ordinators**

  We consulted our teacher group throughout the development of *Power Maths* to ensure we are meeting their real needs in the classroom.

What is *Power Maths*?

Created especially for UK primary schools, and aligned with the new National Curriculum, *Power Maths* is a whole-class, textbook-based mastery resource that empowers every child to understand and succeed. *Power Maths* rejects the notion that some people simply 'can't do' maths. Instead, it develops growth mindsets and encourages hard work, practice and a willingness to see mistakes as learning tools.

Best practice consistently shows that mastery of small, cumulative steps builds a solid foundation of deep mathematical understanding. *Power Maths* combines interactive teaching tools, high-quality textbooks and continuing professional development (CPD) to help you equip children with a deep and long-lasting understanding. Based on extensive evidence, and developed in partnership with practising teachers, *Power Maths* ensures that it meets the needs of children in the UK.

Power Maths and Mastery

Power Maths makes mastery practical and achievable by providing the structures, pathways, content, tools and support you need to make it happen in your classroom.

To develop mastery in maths, children must be enabled to acquire a deep understanding of maths concepts, structures and procedures, step by step. Complex mathematical concepts are built on simpler conceptual components and when children understand every step in the learning sequence, maths becomes transparent and makes logical sense. Interactive lessons establish deep understanding in small steps, as well as effortless fluency in key facts such as tables and number bonds. The whole class works on the same content and no child is left behind.

Power Maths

- ⚡ Builds every concept in small, progressive steps
- ⚡ Is built with interactive, whole-class teaching in mind
- ⚡ Provides the tools you need to develop growth mindsets
- ⚡ Helps you check understanding and ensure that every child is keeping up
- ⚡ Establishes core elements such as intelligent practice and reflection

The *Power Maths* approach

Everyone can!

Founded on the conviction that every child can achieve, *Power Maths* enables children to build number fluency, confidence and understanding, step by step.

Child-centred learning

Children master concepts one step at a time in lessons that embrace a concrete-pictorial-abstract (C-P-A) approach, avoid overload, build on prior learning and help them see patterns and connections. Same-day intervention ensures sustained progress.

Continuing professional development

Embedded teacher support and development offer every teacher the opportunity to continually improve their subject knowledge and manage whole-class teaching for mastery.

Whole-class teaching

An interactive, whole-class teaching model encourages thinking and precise mathematical language and allows children to deepen their understanding as far as they can.

What's different in the new edition?

If you have previously used the first editions of *Power Maths*, you might be interested to know how this edition is different. All of the improvements described below are based on feedback from *Power Maths* customers.

Changes to units and the progression

⚡ The order of units has been slightly adjusted, creating closer alignment between adjacent year groups, which will be useful for mixed age teaching.

⚡ The flow of lessons has been improved within units to optimise the pace of the progression and build in more recap where needed. For key topics, the sequence of lessons gives more opportunities to build up a solid base of understanding. Other units have fewer lessons than before, where appropriate, making it possible to fit in all the content.

⚡ Overall, the lessons put more focus on the most essential content for that year, with less time given to non-statutory content.

⚡ The progression of lessons matches the steps in the new White Rose Maths schemes of learning.

Lesson resources

⚡ There is a Quick recap for each lesson in the Teacher Guide, which offers an alternative lesson starter to the Power Up for cases where you feel it would be more beneficial to surface prerequisite learning than general number fluency.

⚡ In the **Discover** and **Share** sections there is now more of a progression from 1 a) to 1 b). Whereas before, 1 b) was mainly designed as a separate question, now 1 a) leads directly into 1 b). This means that there is an improved whole-class flow, and also an opportunity to focus on the logic and skills in more detail. As a teacher, you will be using 1 a) to lead the class into the thinking, then 1 b) to mould that thinking into the core new learning of the lesson.

⚡ In the **Share** section, for KS1 in particular, the number of different models and representations has been reduced, to support the clarity of thinking prompted by the flow from 1 a) into 1 b).

⚡ More fluency questions have been built into the guided and independent practice.

⚡ Pupil pages are as easy as possible for children to access independently. The pages are less full where this supports greater focus on key ideas and instructions. Also, more freedom is offered around answer format, with fewer boxes scaffolding children's responses; squared paper backgrounds are used in the Practice Books where appropriate. Artwork has also been revisited to ensure the highest standards of accessibility.

New components

480 Individual Practice Games are available in *ActiveLearn* for practising key facts and skills in Years 1 to 6. These are designed in an arcade style, to feel like fun games that children would choose to play outside school. They can be accessed via the Pupil World for homework or additional practice in school – and children can earn rewards. There are Support, Core and Extend levels to allocate, with Activity Reporting available for the teacher. There is a Quick Guide on *ActiveLearn* and you can use the Help area for support in setting up child accounts.

There is also a new set of lesson video resources on the Professional Development tile, designed for in-school training in 10- to 20-minute bursts. For each part of the *Power Maths* lesson sequence, there is a slide deck with embedded video, which will facilitate discussions about how you can take your *Power Maths* teaching to the next level.

Your *Power Maths* resources

Pupil Textbooks

Discover, **Share** and **Think together** sections promote discussion and introduce mathematical ideas logically, so that children understand more easily.

Using a Concrete-Pictorial-Abstract approach, clear mathematical models help children to make connections and grasp concepts.

Appealing scenarios stimulate curiosity, helping children to identify the maths problem and discover patterns and relationships for themselves.

Friendly, supportive characters help children develop a growth mindset by prompting them to think, reason and reflect.

To help you teach for mastery, *Power Maths* comprises a variety of high-quality resources.

The coherent *Power Maths* lesson structure carries through into the vibrant, high-quality textbooks. Setting out the core learning objectives for each class, the lesson structure follows a carefully mapped journey through the curriculum and supports children on their journey to deeper understanding.

Pupil Practice Books

The Practice Books offer just the right amount of intelligent practice for children to complete independently in the final section of each lesson.

Practice questions are finely tuned to move children forward in their thinking and to reveal misconceptions.

The practice questions are for everyone – each question varies one small element to move children on in their thinking.

Calculations are connected so that children think about the underlying concept.

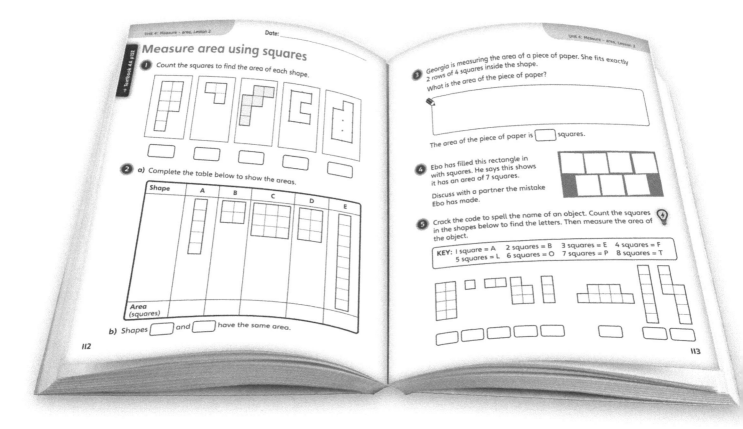

CHALLENGE questions allow children to delve deeper into a concept.

The *Power Maths* characters support and encourage children to think and work in different ways.

Reflect questions reveal the depth of each child's understanding before they move on.

 Think differently questions encourage children to use reasoning as well as their mathematical knowledge to reach a solution.

Online subscription

The online subscription will give you access to additional resources and answers from the Textbook and Practice Book.

eTextbooks

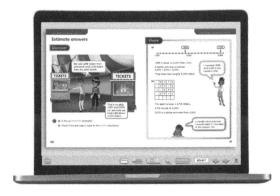

Digital versions of *Power Maths* Textbooks allow class groups to share and discuss questions, solutions and strategies. They allow you to project key structures and representations at the front of the class, to ensure all children are focusing on the same concept.

Teaching tools

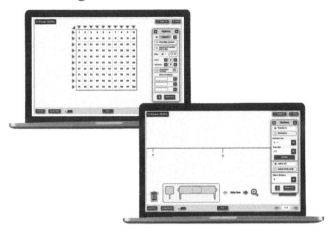

Here you will find interactive versions of key *Power Maths* structures and representations.

Power Ups

Use this series of daily activities to promote and check number fluency.

Online versions of Teacher Guide pages

PDF pages give support at both unit and lesson levels. You will also find help with key strategies and templates for tracking progress.

Unit videos

Watch the professional development videos at the start of each unit to help you teach with confidence. The videos explore common misconceptions in the unit, and include intervention suggestions as well as suggestions on what to look out for when assessing mastery in your students.

End of unit Strengthen and Deepen materials

The Strengthen activity at the end of every unit addresses a key misconception and can be used to support children who need it. The Deepen activities are designed to be low ceiling/high threshold and will challenge those children who can understand more deeply. These resources will help you ensure that every child understands and will help you keep the class moving forward together. These printable activities provide an optional resource bank for use after the assessment stage.

Individual Practice Games

These enjoyable games can be used at home or at school to embed key number skills (see page 6).

Professional Development videos and slides

These slides and videos of *Power Maths* lessons can be used for ongoing training in short bursts or to support new staff.

The *Power Maths* teaching model

At the heart of *Power Maths* is a clearly structured teaching and learning process that helps you make certain that every child masters each maths concept securely and deeply. For each year group, the curriculum is broken down into core concepts, taught in units. A unit divides into smaller learning steps – lessons. Step by step, strong foundations of cumulative knowledge and understanding are built.

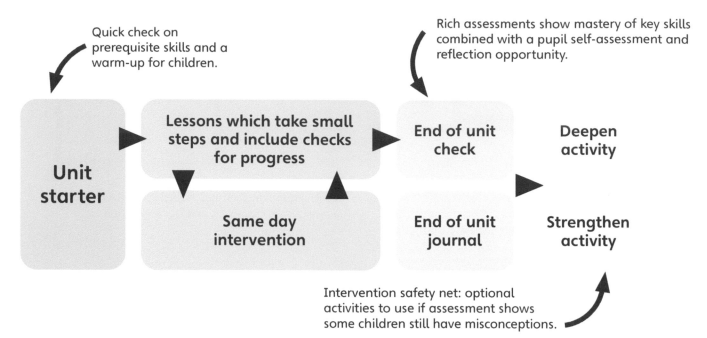

Quick check on prerequisite skills and a warm-up for children.

Rich assessments show mastery of key skills combined with a pupil self-assessment and reflection opportunity.

Unit starter → **Lessons which take small steps and include checks for progress** → **End of unit check** → **Deepen activity**

Same day intervention — **End of unit journal** — **Strengthen activity**

Intervention safety net: optional activities to use if assessment shows some children still have misconceptions.

Unit starter

Each unit begins with a unit starter, which introduces the learning context along with key mathematical vocabulary and structures and representations.

- The Textbooks include a check on readiness and a warm-up task for children to complete.
- Your Teacher Guide gives support right from the start on important structures and representations, mathematical language, common misconceptions and intervention strategies.
- Unit-specific videos develop your subject knowledge and insights so you feel confident and fully equipped to teach each new unit. These are available via the online subscription.

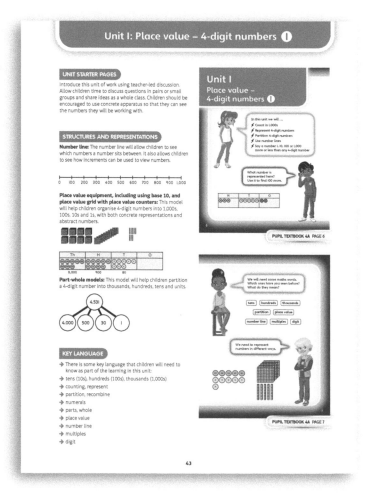

Lesson

Once a unit has been introduced, it is time to start teaching the series of lessons.

- Each lesson is scaffolded with Textbook and Practice Book activities and begins with a Power Up activity (available via online subscription) or the Quick recap activity in the Teacher Guide (see page 15).
- *Power Maths* identifies lesson by lesson what concepts are to be taught.
- Your Teacher Guide offers lots of support for you to get the most from every child in every lesson. As well as highlighting key points, tricky areas and how to handle them, you will also find question prompts to check on understanding and clarification on why particular activities and questions are used.

Same-day intervention

Same-day interventions are vital in order to keep the class progressing together. This can be during the lesson as well as afterwards (see page 27). Therefore, *Power Maths* provides plenty of support throughout the journey.

- Intervention is focused on keeping up now, not catching up later, so interventions should happen as soon as they are needed.
- Practice section questions are designed to bring misconceptions to the surface, allowing you to identify these easily as you circulate during independent practice time.
- Child-friendly assessment questions in the Teacher Guide help you identify easily which children need to strengthen their understanding.

End of unit check and journal

For each unit, the End of unit check in the Textbook lets you see which children have mastered the key concepts, which children have not and where their misconceptions lie. The Practice Books also include an End of unit journal in which children can reflect on what they have learned. Each unit also offers Strengthen and Deepen activities, available via the online subscription.

The Teacher Guide offers different ways of managing the End of unit assessments as well as giving support with handling misconceptions.

The End of unit check presents multiple-choice questions. Children think about their answer, decide on a solution and explain their choice.

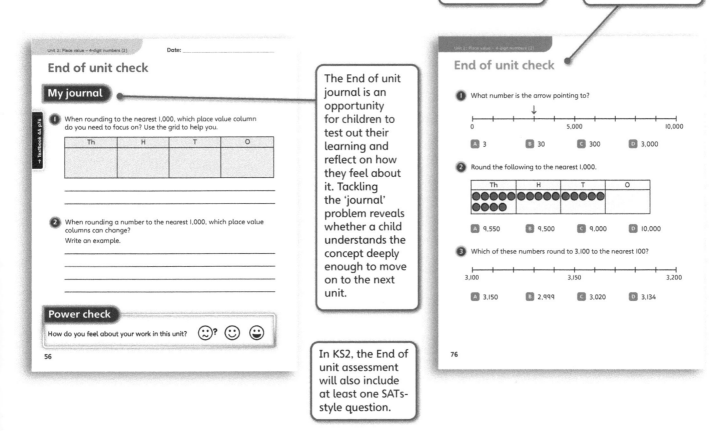

The End of unit journal is an opportunity for children to test out their learning and reflect on how they feel about it. Tackling the 'journal' problem reveals whether a child understands the concept deeply enough to move on to the next unit.

In KS2, the End of unit assessment will also include at least one SATs-style question.

The *Power Maths* lesson sequence

At the heart of *Power Maths* is a unique lesson sequence designed to empower children to understand core concepts and grow in confidence. Embracing the National Centre for Excellence in the Teaching of Mathematics' (NCETM's) definition of mastery, the sequence guides and shapes every *Power Maths* lesson you teach.

Flexibility is built into the *Power Maths* programme so there is no one-to-one mapping of lessons and concepts and you can pace your teaching according to your class. While some children will need to spend longer on a particular concept (through interventions or additional lessons), others will reach deeper levels of understanding. However, it is important that the class moves forward together through the termly schedules.

Power Up ⏱ 5 minutes

Each lesson begins with a Power Up activity (available via the online subscription) which supports fluency in key number facts.

The whole-class approach depends on fluency, so the Power Up is a powerful and essential activity.

The Quick recap is an alternative starter, for when you think some or all children would benefit more from revisiting pre-requisite work (see page 15).

TOP TIP
If the class is struggling with the task, revisit it later and check understanding.

Power Ups reinforce the two key things that are essential for success: times-tables and number bonds.

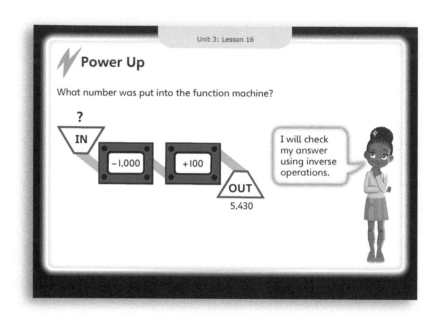

Discover ⏱ 10 minutes

A practical, real-life problem arouses curiosity. Children find the maths through story telling.

A real-life scenario is provided for the **Discover** section but feel free to build upon these with your own examples that are more relevant to your class, or get creative with the context.

TOP TIP
Discover works best when run at tables, in pairs with concrete objects.

Question ❶ a) tackles the key concept and question ❶ b) digs a little deeper. Children have time to explore, play and discuss possible strategies.

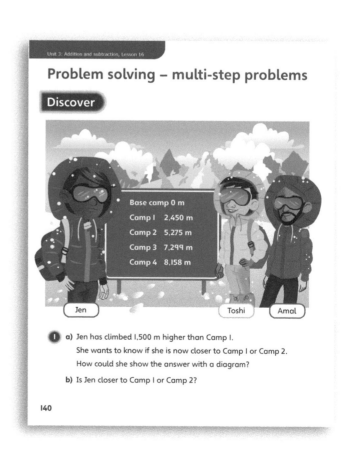

Share 10 minutes

Teacher-led, this interactive section follows the **Discover** activity and highlights the variety of methods that can be used to solve a single problem.

TOP TIP

Pairs sharing a textbook is a great format for **Share**!

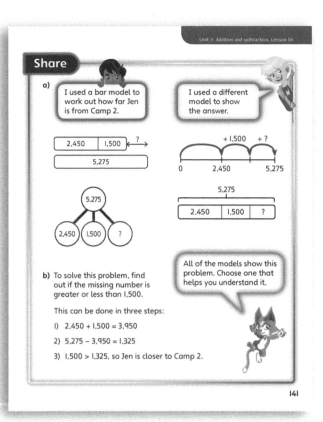

Your Teacher Guide gives target questions for children. The online toolkit provides interactive structures and representations to link concrete and pictorial to abstract concepts.

Bring children to the front to share and celebrate their solutions and strategies.

Think together

 10 minutes

Children work in groups on the carpet or at tables, using their textbooks or eBooks.

TOP TIP

Make sure children have mini whiteboards or pads to write on if they are not at their tables.

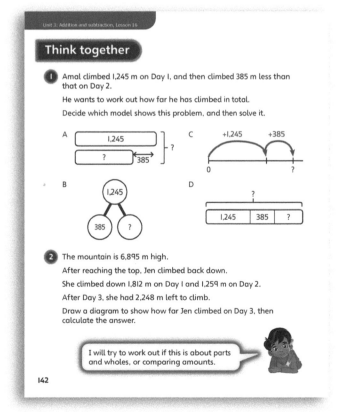

Using the Teacher Guide, model question ❶ for your class.

Question ❷ is less structured. Children will need to think together in their groups, then discuss their methods and solutions as a class.

In question ❸ children try working out the answer independently. The openness of the **Challenge** question helps to check depth of understanding.

Practice ⏱ 15 minutes

Using their Practice Books, children work independently while you circulate and check on progress.

Questions follow small steps of progression to deepen learning.

TOP TIP
Some children could work separately with a teacher or assistant.

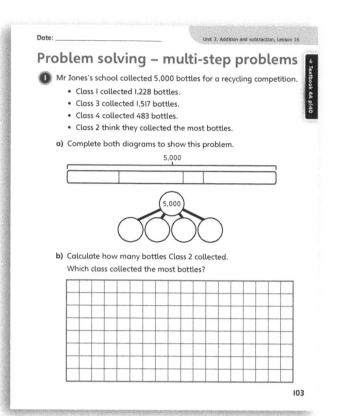

Date: _____

Problem solving – multi-step problems

→ Textbook 4A p140

1 Mr Jones's school collected 5,000 bottles for a recycling competition.
- Class 1 collected 1,228 bottles.
- Class 3 collected 1,517 bottles.
- Class 4 collected 483 bottles.
- Class 2 think they collected the most bottles.

a) Complete both diagrams to show this problem.

5,000

5,000

b) Calculate how many bottles Class 2 collected. Which class collected the most bottles?

103

Are some children struggling? If so, work with them as a group, using mathematical structures and representations to support understanding as necessary.

There are no set routines: for real understanding, children need to think about the problem in different ways.

Reflect ⏱ 5 minutes

'Spot the mistake' questions are great for checking misconceptions.

The **Reflect** section is your opportunity to check how deeply children understand the target concept.

4 Write a story problem to match the diagram. **CHALLENGE**

Class 1 [   ]   950
Class 2 [   ]   ⎤
Class 3 [1,900   ] ⎦ 4,000

_____
_____
_____
_____
_____
_____
_____
_____

Reflect

When I draw a bar model to help me solve a problem, I decide how many bars I need to draw by _____
_____
_____
_____

105

The Practice Books use various approaches to check that children have fully understood each concept.

Looking like they understand is not enough! It is essential that children can show they have grasped the concept.

Using the *Power Maths* Teacher Guide

Think of your Teacher Guides as *Power Maths* handbooks that will guide, support and inspire your day-to-day teaching. Clear and concise, and illustrated with helpful examples, your Teacher Guides will help you make the best possible use of every individual lesson. They also provide wrap-around professional development, enhancing your own subject knowledge and helping you to grow in confidence about moving your children forward together.

There is a Teacher Guide per year group for every term, with unit and lesson level guidance and support.

Never feel stuck! You will find ideas for introducing every unit and lesson and questions to encourage teacher reflection before and after each lesson.

Tips and advice on key elements such as C-P-A approaches, misconceptions, language, modelling growth mindsets and same day intervention.

Annotations for every Textbook and Practice Book page, providing prompts for key questions to ask to expose understanding and explanations as to why key questions have been chosen.

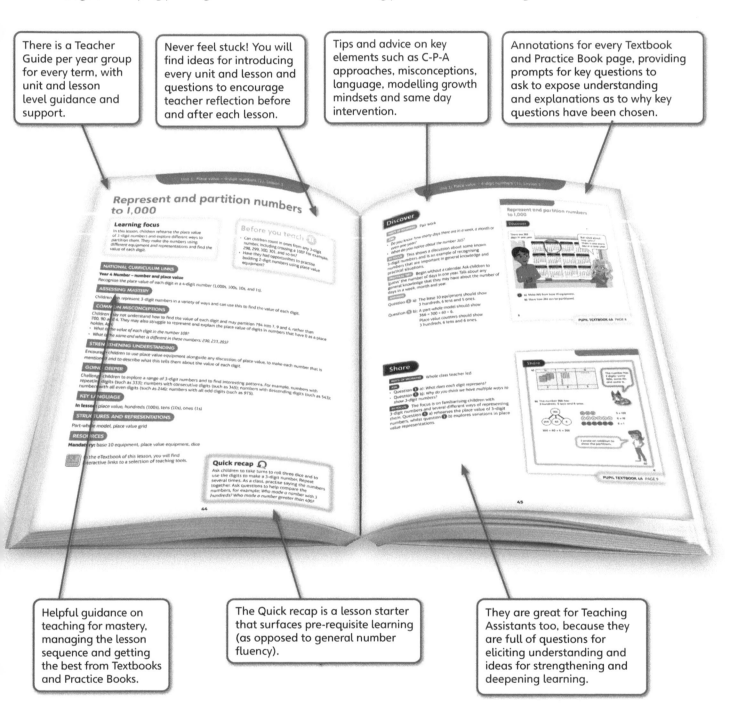

Helpful guidance on teaching for mastery, managing the lesson sequence and getting the best from Textbooks and Practice Books.

The Quick recap is a lesson starter that surfaces pre-requisite learning (as opposed to general number fluency).

They are great for Teaching Assistants too, because they are full of questions for eliciting understanding and ideas for strengthening and deepening learning.

At the end of each unit, your Teacher Guide helps you identify who has fully grasped the concept, who has not and how to move every child forward. This is covered later in the Assessment strategies section.

Power Maths Year 4, yearly overview

Textbook	Strand	Unit		Number of lessons
Textbook A / Practice Workbook A (Term 1)	Number – number and place value	1	Place value – 4-digit numbers (1)	8
	Number – number and place value	2	Place value – 4-digit numbers (2)	8
	Number – addition and subtraction	3	Addition and subtraction	16
	Measurement	4	Measure – area	5
	Number – multiplication and division	5	Multiplication and division (1)	12
Textbook B / Practice Workbook B (Term 2)	Number – multiplication and division	6	Multiplication and division (2)	16
	Measurement	7	Length and perimeter	6
	Number – fractions	8	Fractions (1)	9
	Number – fractions	9	Fractions (2)	8
	Number – fractions (including decimals and percentages	10	Decimals (1)	12
Textbook C / Practice Workbook C (Term 3)	Number – fractions (including decimals and percentages	11	Decimals (2)	7
	Measurement	12	Money	6
	Measurement	13	Time	5
	Geometry – properties of shapes	14	Geometry – angles and 2D shapes	8
	Statistics	15	Statistics	6
	Geometry – position and direction	16	Geometry – position and direction	6

Power Maths Year 4, Textbook 4A (Term I) overview

Strand	Unit		Lesson number	Lesson title	NC Objective 1	NC Objective 2
Number – number and place value	Unit 1	Place value – 4-digit numbers (1)	1	Represent and partition numbers to 1,000	Recognise the place value of each digit in a four-digit number (1,000s, 100s, 10s, and 1s)	
Number – number and place value	Unit 1	Place value – 4-digit numbers (1)	2	Number line to 1,000	Recognise the place value of each digit in a four-digit number (1,000s, 100s, 10s, and 1s)	
Number – number and place value	Unit 1	Place value – 4-digit numbers (1)	3	Multiples of 1,000	Count in multiples of 6, 7, 9, 25 and 1,000	
Number – number and place value	Unit 1	Place value – 4-digit numbers (1)	4	4-digit numbers	Identify, represent and estimate numbers using different representations	
Number – number and place value	Unit 1	Place value – 4-digit numbers (1)	5	Partition 4-digit numbers flexibly	Recognise the place value of each digit in a four-digit number (1,000s, 100s, 10s, and 1s)	
Number – number and place value	Unit 1	Place value – 4-digit numbers (1)	6	Partition 4-digit numbers flexibly	Recognise the place value of each digit in a four-digit number (1,000s, 100s, 10s, and 1s)	Identify, represent and estimate numbers using different representations
Number – number and place value	Unit 1	Place value – 4-digit numbers (1)	7	1, 10, 100, 1,000 more or less	Find 1,000 more or less than a given number	Count from 0 in multiples of 4, 8, 50 and 100; find 10 or 100 more or less than a given number
Number – number and place value	Unit 1	Place value – 4-digit numbers (1)	8	1,000s, 100s, 10s and 1s	Recognise the place value of each digit in a four-digit number (1,000s, 100s, 10s, and 1s)	Identify, represent and estimate numbers using different representations
Number – number and place value	Unit 2	Place value – 4-digit numbers (2)	1	Number line to 10,000	Identify, represent and estimate numbers using different representations	Recognise the place value of each digit in a four-digit number (1,000s, 100s, 10s, and 1s)
Number – number and place value	Unit 2	Place value – 4-digit numbers (2)	2	Between two multiples	Recognise the place value of each digit in a four-digit number (1,000s, 100s, 10s, and 1s)	Count in multiples of 6, 7, 9, 25 and 1000
Number – number and place value	Unit 2	Place value – 4-digit numbers (2)	3	Estimate on a number line to 10,000	Order and compare numbers beyond 1,000	Identify, represent and estimate numbers using different representations
Number – number and place value	Unit 2	Place value – 4-digit numbers (2)	4	Compare and order numbers to 10,000	Order and compare numbers beyond 1,000	Identify, represent and estimate numbers using different representations
Number – number and place value	Unit 2	Place value – 4-digit numbers (2)	5	Round to the nearest 1,000	Round any number to the nearest 10, 100 or 1,000	
Number – number and place value	Unit 2	Place value – 4-digit numbers (2)	6	Round to the nearest 100	Round any number to the nearest 10, 100 or 1,000	
Number – number and place value	Unit 2	Place value – 4-digit numbers (2)	7	Round to the nearest 10	Round any number to the nearest 10, 100 or 1,000	
Number – number and place value	Unit 2	Place value – 4-digit numbers (2)	8	Round to the nearest 1,000, 100 or 10	Round any number to the nearest 10, 100 or 1,000	
Number – addition and subtraction	Unit 3	Addition and subtraction	1	Add and subtract 1s, 10s, 100s, 1,000s	Add and subtract numbers with up to 4 digits using the formal written methods of columnar addition and subtraction where appropriate	Solve number and practical problems that involve all of the above and with increasingly large positive numbers
Number – addition and subtraction	Unit 3	Addition and subtraction	2	Add two 4-digit numbers – one exchange	Add and subtract numbers with up to 4 digits using the formal written methods of columnar addition and subtraction where appropriate	
Number – addition and subtraction	Unit 3	Addition and subtraction	3	Add two 4-digit numbers – one exchange	Add and subtract numbers with up to 4 digits using the formal written methods of columnar addition and subtraction where appropriate	

Strand	Unit		Lesson number	Lesson title	NC Objective 1	NC Objective 2
Number – addition and subtraction	Unit 3	Addition and subtraction	4	Add with more than one exchange	Add and subtract numbers with up to 4 digits using the formal written methods of columnar addition and subtraction where appropriate	
Number – addition and subtraction	Unit 3	Addition and subtraction	5	Subtract two 4-digit numbers	Add and subtract numbers with up to 4 digits using the formal written methods of columnar addition and subtraction where appropriate	
Number – addition and subtraction	Unit 3	Addition and subtraction	6	Subtract two 4-digit numbers – one exchange	Add and subtract numbers with up to 4 digits using the formal written methods of columnar addition and subtraction where appropriate	
Number – addition and subtraction	Unit 3	Addition and subtraction	7	Subtract two 4-digit numbers – more than one exchange	Add and subtract numbers with up to 4 digits using the formal written methods of columnar addition and subtraction where appropriate	
Number – addition and subtraction	Unit 3	Addition and subtraction	8	Exchange across two columns	Add and subtract numbers with up to 4 digits using the formal written methods of columnar addition and subtraction where appropriate	
Number – addition and subtraction	Unit 3	Addition and subtraction	9	Efficient methods	Estimate and use inverse operations to check answers to a calculation	Add and subtract numbers with up to 4 digits using the formal written methods of columnar addition and subtraction where appropriate
Number – addition and subtraction	Unit 3	Addition and subtraction	10	Equivalent difference	Estimate and use inverse operations to check answers to a calculation	
Number – addition and subtraction	Unit 3	Addition and subtraction	11	Estimate answers	Estimate and use inverse operations to check answers to a calculation	
Number – addition and subtraction	Unit 3	Addition and subtraction	12	Check strategies	Estimate and use inverse operations to check answers to a calculation	
Number – addition and subtraction	Unit 3	Addition and subtraction	13	Problem solving – one step	Solve addition and subtraction two-step problems in contexts, deciding which operations and methods to use and why	
Number – addition and subtraction	Unit 3	Addition and subtraction	14	Problem solving – comparison	Solve addition and subtraction two-step problems in contexts, deciding which operations and methods to use and why	
Number – addition and subtraction	Unit 3	Addition and subtraction	15	Problem solving – two steps	Solve addition and subtraction two-step problems in contexts, deciding which operations and methods to use and why	
Number – addition and subtraction	Unit 3	Addition and subtraction	16	Problem solving – multi-step problems	Solve addition and subtraction two-step problems in contexts, deciding which operations and methods to use and why	
Measurement	Unit 4	Measure – area	1	What is area?	Find the area of rectilinear shapes by counting squares	
Measurement	Unit 4	Measure – area	2	Measure area using squares	Find the area of rectilinear shapes by counting squares	
Measurement	Unit 4	Measure – area	3	Count squares	Find the area of rectilinear shapes by counting squares	
Measurement	Unit 4	Measure – area	4	Make shapes	Find the area of rectilinear shapes by counting squares	
Measurement	Unit 4	Measure – area	5	Compare area	Estimate, compare and calculate different measures, including money in pounds and pence	

Strand	Unit		Lesson number	Lesson title	NC Objective 1	NC Objective 2
Number – multiplication and division	Unit 5	Multiplication and division (1)	1	Multiples of 3	Recall multiplication and division facts for multiplication tables up to 12 × 12	
Number – multiplication and division	Unit 5	Multiplication and division (1)	2	Multiply and divide by 6	Recall multiplication and division facts for multiplication tables up to 12 × 12	
Number – multiplication and division	Unit 5	Multiplication and division (1)	3	6 times-table and division facts	Recall multiplication and division facts for multiplication tables up to 12 × 12	
Number – multiplication and division	Unit 5	Multiplication and division (1)	4	Multiply and divide by 9	Recall multiplication and division facts for multiplication tables up to 12 × 12	
Number – multiplication and division	Unit 5	Multiplication and division (1)	5	9 times-table and division facts	Recall multiplication and division facts for multiplication tables up to 12 × 12	
Number – multiplication and division	Unit 5	Multiplication and division (1)	6	The 3, 6 and 9 times-tables	Recall multiplication and division facts for multiplication tables up to 12 × 12	
Number – multiplication and division	Unit 5	Multiplication and division (1)	7	Multiply and divide by 7	Recall multiplication and division facts for multiplication tables up to 12 × 12	
Number – multiplication and division	Unit 5	Multiplication and division (1)	8	7 times-table and division facts	Recall multiplication and division facts for multiplication tables up to 12 × 12	
Number – multiplication and division	Unit 5	Multiplication and division (1)	9	11 and 12 times-tables and division facts	Recall multiplication and division facts for multiplication tables up to 12 × 12	
Number – multiplication and division	Unit 5	Multiplication and division (1)	10	Multiply by 1 and 0	Use place value, known and derived facts to multiply and divide mentally, including: multiplying by 0 and 1; dividing by 1; multiplying together three numbers	
Number – multiplication and division	Unit 5	Multiplication and division (1)	11	Divide by 1 and itself	Use place value, known and derived facts to multiply and divide mentally, including: multiplying by 0 and 1; dividing by 1; multiplying together three numbers	
Number – multiplication and division	Unit 5	Multiplication and division (1)	12	Multiply three numbers	Use place value, known and derived facts to multiply and divide mentally, including: multiplying by 0 and 1; dividing by 1; multiplying together three numbers	

Mindset: an introduction

Global research and best practice deliver the same message: learning is greatly affected by what learners perceive they can or cannot do. What is more, it is also shaped by what their parents, carers and teachers perceive they can do. Mindset – the thinking that determines our beliefs and behaviours – therefore has a fundamental impact on teaching and learning.

Everyone can!

Power Maths and mastery methods focus on the distinction between 'fixed' and 'growth' mindsets (Dweck, 2007).[1] Those with a fixed mindset believe that their basic qualities (for example, intelligence, talent and ability to learn) are pre-wired or fixed: 'If you have a talent for maths, you will succeed at it. If not, too bad!' By contrast, those with a growth mindset believe that hard work, effort and commitment drive success and that 'smart' is not something you are or are not, but something you become. In short, everyone can do maths!

Key mindset strategies

A growth mindset needs to be actively nurtured and developed. *Power Maths* offers some key strategies for fostering healthy growth mindsets in your classroom.

It is okay to get it wrong

Mistakes are valuable opportunities to re-think and understand more deeply. Learning is richer when children and teachers alike focus on spotting and sharing mistakes as well as solutions.

Praise hard work

Praise is a great motivator, and by focusing on praising effort and learning rather than success, children will be more willing to try harder, take risks and persist for longer.

Mind your language!

The language we use around learners has a profound effect on their mindsets. Make a habit of using growth phrases, such as, 'Everyone can!', 'Mistakes can help you learn' and 'Just try for a little longer'. The king of them all is one little word, 'yet'... I can't solve this...yet!' Encourage parents and carers to use the right language too.

Build in opportunities for success

The step-by-small-step approach enables children to enjoy the experience of success. In addition, avoid ability grouping and encourage every child to answer questions and explain or demonstrate their methods to others.

The *Power Maths* characters

The *Power Maths* characters model the traits of growth mindset learners and encourage resilience by prompting and questioning children as they work. Appearing frequently in the Textbooks and Practice Books, they are your allies in teaching and discussion, helping to model methods, alternatives and misconceptions, and to pose questions. They encourage and support your children, too: they are all hardworking, enthusiastic and unafraid of making and talking about mistakes.

Meet the team!

Creative Flo is open-minded and sometimes indecisive. She likes to think differently and come up with a variety of methods or ideas.

Determined Dexter is resolute, resilient and systematic. He concentrates hard, always tries his best and he'll never give up – even though he doesn't always choose the most efficient methods!

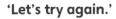

'Let's try again.'

'Mistakes are cool!'

'Have I found all of the solutions?'

'Let's try it this way...'

'Can we do it differently?'

'I've got another way of doing this!'

'I'm going to try this!'

'I know how to do that!'

'Want to share my ideas?'

Curious Ash is eager, interested and inquisitive, and he loves solving puzzles and problems. Ash asks lots of questions but sometimes gets distracted.

'What if we tried this...?'

'I wonder...'

'Is there a pattern here?'

Sparks the Cat

Miaow!

Brave Astrid is confident, willing to take risks and unafraid of failure. She's never scared to jump straight into a problem or question, and although she often makes simple mistakes, she's happy to talk them through with others.

Mathematical language

Traditionally, we in the UK have tended to try simplifying mathematical language to make it easier for young children to understand. By contrast, evidence and experience show that by diluting the correct language, we actually mask concepts and meanings for children. We then wonder why they are confused by new and different terminology later down the line! *Power Maths* is not afraid of 'hard' words and avoids placing any barriers between children and their understanding of mathematical concepts. As a result, we need to be deliberate, precise and thorough in building every child's understanding of the language of maths. Throughout the Teacher Guides you will find support and guidance on how to deliver this, as well as individual explanations throughout the pupil Textbooks.

Use the following key strategies to build children's mathematical vocabulary, understanding and confidence.

Precise and consistent

Everyone in the classroom should use the correct mathematical terms in full, every time. For example, refer to 'equal parts', not 'parts'. Used consistently, precise maths language will be a familiar and non-threatening part of children's everyday experience.

Full sentences

Teachers and children alike need to use full sentences to explain or respond. When children use complete sentences, it both reveals their understanding and embeds their knowledge.

Stem sentences

These important sentences help children express mathematical concepts accurately, and are used throughout the *Power Maths* books. Encourage children to repeat them frequently, whether working independently or with others. Examples of stem sentences are:

'4 is a part, 5 is a part, 9 is the whole.'

'There are .... groups. There are .... in each group.'

Key vocabulary

The unit starters highlight essential vocabulary for every lesson. In the pupil books, characters flag new terminology and the Teacher Guide lists important mathematical language for every unit and lesson. New terms are never introduced without a clear explanation.

Mathematical signs

Mathematical signs are used early on so that children quickly become familiar with them and their meaning. Often, the *Power Maths* characters will highlight the connection between language and particular signs.

The role of talk and discussion

When children learn to talk purposefully together about maths, barriers of fear and anxiety are broken down and they grow in confidence, skills and understanding. Building a healthy culture of 'maths talk' empowers their learning from day one.

Explanation and discussion are integral to the *Power Maths* structure, so by simply following the books your lessons will stimulate structured talk. The following key 'maths talk' strategies will help you strengthen that culture and ensure that every child is included.

Sentences, not words

Encourage children to use full sentences when reasoning, explaining or discussing maths. This helps both speaker and listeners to clarify their own understanding. It also reveals whether or not the speaker truly understands, enabling you to address misconceptions as they arise.

Working together

Working with others in pairs, groups or as a whole class is a great way to support maths talk and discussion. Use different group structures to add variety and challenge. For example, children could take timed turns for talking, work independently alongside a 'discussion buddy', or perhaps play different *Power Maths* character roles within their group.

Think first – then talk

Provide clear opportunities within each lesson for children to think and reflect, so that their talk is purposeful, relevant and focused.

Give every child a voice

Where the 'hands up' model allows only the more confident child to shine, *Power Maths* involves everyone. Make sure that no child dominates and that even the shyest child is encouraged to contribute – and praised when they do.

Assessment strategies

Teaching for mastery demands that you are confident about what each child knows and where their misconceptions lie; therefore, practical and effective assessment is vitally important.

Formative assessment within lessons

The **Think together** section will often reveal any confusions or insecurities; try ironing these out by doing the first **Think together** question as a class. For children who continue to struggle, you or your Teaching Assistant should provide support and enable them to move on.

▶

Performance in practice can be very revealing: check Practice Books and listen out both during and after practice to identify misconceptions.

▶

The **Reflect** section is designed to check on the all-important depth of understanding. Be sure to review how the children performed in this final stage before you teach the next lesson.

End of unit check – Textbook

Each unit concludes with a summative check to help you assess quickly and clearly each child's understanding, fluency, reasoning and problem solving skills. Your Teacher Guide will suggest ideal ways of organising a given activity and offer advice and commentary on what children's responses mean. For example, 'What misconception does this reveal?'; 'How can you reinforce this particular concept?'

Assessment with young children should always be an enjoyable activity, so avoid one-to-one individual assessments, which they may find threatening or scary. If you prefer, the End of unit check can be carried out as a whole-class group using whiteboards and Practice Books.

End of unit check – Practice Book

The Practice Book contains further opportunities for assessment, and can be completed by children independently whilst you are carrying out diagnostic assessment with small groups. Your Teacher Guide will advise you on what to do if children struggle to articulate an explanation – or perhaps encourage you to write down something they have explained well. It will also offer insights into children's answers and their implications for next learning steps. It is split into three main sections, outlined below.

My journal is designed to allow children to show their depth of understanding of the unit. It can also serve as a way of checking that children have grasped key mathematical vocabulary. The question children should answer is first presented in the Textbook in the Think! section. This provides an opportunity for you to discuss the question first as a class to ensure children have understood their task. Children should have some time to think about how they want to answer the question, and you could ask them to talk to a partner about their ideas. Then children should write their answer in their Practice Book, using the word bank provided to help them with vocabulary.

The **Power check** allows pupils to self-assess their level of confidence on the topic by colouring in different smiley faces. You may want to introduce the faces as follows:

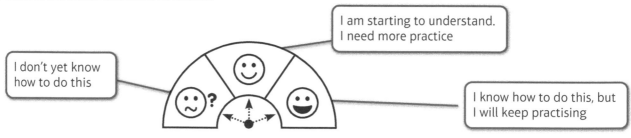

Each unit ends with either a Power play or a Power puzzle. This is an activity, puzzle or game that allows children to use their new knowledge in a fun, informal way.

Progress Tests

There are *Power Maths* Progress Tests for each half term and at the end of the year, including an Arithmetic test and Reasoning test in each case. You can enter results in the online markbook to track and analyse results and see the average for all schools' results. The tests use a 6-step scale to show results against age-related expectation.

How to ask diagnostic questions

The diagnostic questions provided in children's Practice Books are carefully structured to identify both understanding and misconceptions (if children answer in a particular way, you will know why). The simple procedure below may be helpful:

Ask the question, offering the selection of answers provided.

▼

Children take time to think about their response.

▼

Each child selects an answer and shares their reasoning with the group.

▼

Give minimal and neutral feedback (for example, 'That's interesting', or 'Okay').

▼

Ask, 'Why did you choose that answer?', then offer an opportunity to change their mind by providing one correct and one incorrect answer.

▼

Note which children responded and reasoned correctly first time and everyone's final choices.

▼

Reflect that together, we can get the right answer.

Keeping the class together

Traditionally, children who learn quickly have been accelerated through the curriculum. As a consequence, their learning may be superficial and will lack the many benefits of enabling children to learn with and from each other.

By contrast, *Power Maths'* mastery approach values real understanding and richer, deeper learning above speed. It sees all children learning the same concept in small, cumulative steps, each finding and mastering challenge at their own level. Remember that when you teach for mastery, EVERYONE can do maths! Those who grasp a concept easily have time to explore and understand that concept at a deeper level. The whole class therefore moves through the curriculum at broadly the same pace via individual learning journeys.

For some teachers, the idea that a whole class can move forward together is revolutionary and challenging. However, the evidence of global good practice clearly shows that this approach drives engagement, confidence, motivation and success for all learners, and not just the high flyers. The strategies below will help you keep your class together on their maths journey.

Mix it up

Do not stick to set groups at each table. Every child should be working on the same concept, and mixing up the groupings widens children's opportunities for exploring, discussing and sharing their understanding with others.

Recycling questions

Reuse the Textbook and Practice Book questions with concrete materials to allow children to explore concepts and relationships and deepen their understanding. This strategy is especially useful for reinforcing learning in same-day interventions.

Strengthen at every opportunity

The next lesson in a *Power Maths* sequence always revises and builds on the previous step to help embed learning. These activities provide golden opportunities for individual children to strengthen their learning with the support of Teaching Assistants.

Prepare to be surprised!

Children may grasp a concept quickly or more slowly. The 'fast graspers' won't always be the same individuals, nor does the speed at which a child understands a concept predict their success in maths. Are they struggling or just working more slowly?

Same-day intervention

Since maths competence depends on mastering concepts one by one in a logical progression, it is important that no gaps in understanding are ever left unfilled. Same-day interventions – either within or after a lesson – are a crucial safety net for any child who has not fully made the small step covered that day. In other words, intervention is always about keeping up, not catching up, so that every child has the skills and understanding they need to tackle the next lesson. That means presenting the same problems used in the lesson, with a variety of concrete materials to help children model their solutions.

We offer two intervention strategies below, but you should feel free to choose others if they work better for your class.

Within-lesson intervention

The **Think together** activity will reveal those who are struggling, so when it is time for practice, bring these children together to work with you on the first practice questions. Observe these children carefully, ask questions, encourage them to use concrete models and check that they reach and can demonstrate their understanding.

After-lesson intervention

You might like to use the **Think together** questions to recap the lesson with children who are working behind expectations during assembly time. Teaching Assistants could also work with these children at other convenient points in the school day. Some children may benefit from revisiting work from the same topic in the previous year group. Note also the suggestion for recycling questions from the Textbook and Practice Book with concrete materials on page 26.

The role of practice

Practice plays a pivotal role in the *Power Maths* approach. It takes place in class groups, smaller groups, pairs, and independently, so that children always have the opportunities for thinking as well as the models and support they need to practise meaningfully and with understanding.

Intelligent practice

In *Power Maths*, practice never equates to the simple repetition of a process. Instead we embrace the concept of intelligent practice, in which all children become fluent in maths through varied, frequent and thoughtful practice that deepens and embeds conceptual understanding in a logical, planned sequence. To see the difference, take a look at the following examples.

Traditional practice

- Repetition can be rote – no need for a child to think hard about what they are doing

- Praise may be misplaced

- Does this prove understanding?

Intelligent practice

- Varied methods – concrete, pictorial and abstract

- Equation expressed in different ways, requiring thought and understanding

- Constructive feedback

All practice questions are designed to move children on and reveal misconceptions.

Simple, logical steps build onto earlier learning.

C-P-A runs throughout – different ways of modelling and understanding the same concept.

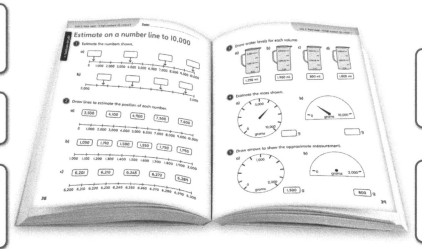

Conceptual variation – children work on different representations of the same maths concept.

Friendly characters offer support and encourage children to try different approaches.

A carefully designed progression

The Practice Books provide just the right amount of intelligent practice for children to complete independently in the final sections of each lesson. It is really important that all children are exposed to the practice questions, and that children are not directed to complete different sections. That is because each question is different and has been designed to challenge children to think about the maths they are doing. The questions become more challenging so children grasping concepts more quickly will start to slow down as they progress. Meanwhile, you have the chance to circulate and spot any misconceptions before they become barriers to further learning.

Homework and the role of parents and carers

While *Power Maths* does not prescribe any particular homework structure, we acknowledge the potential value of practice at home. For example, practising fluency in key facts, such as number bonds and times-tables, is an ideal homework task. You can share the Individual Practice Games for homework (see page 6), or parents and carers could work through uncompleted Practice Book questions with children at either primary stage.

However, it is important to recognise that many parents and carers may themselves lack confidence in maths, and few, if any, will be familiar with mastery methods. A Parents' and Carers' evening that helps them understand the basics of mindsets, mastery and mathematical language is a great way to ensure that children benefit from their homework. It could be a fun opportunity for children to teach their families that everyone can do maths!

Structures and representations

Unlike most other subjects, maths comprises a wide array of abstract concepts – and that is why children and adults so often find it difficult. By taking a concrete-pictorial-abstract (C-P-A) approach, *Power Maths* allows children to tackle concepts in a tangible and more comfortable way.

Non-linear stages

Concrete

Replacing the traditional approach of a teacher working through a problem in front of the class, the concrete stage introduces real objects that children can use to 'do' the maths – any familiar object that a child can manipulate and move to help bring the maths to life. It is important to appreciate, however, that children must always understand the link between models and the objects they represent. For example, children need to first understand that three cakes could be represented by three pretend cakes, and then by three counters or bricks. Frequent practice helps consolidate this essential insight. Although they can be used at any time, good concrete models are an essential first step in understanding.

Pictorial

This stage uses pictorial representations of objects to let children 'see' what particular maths problems look like. It helps them make connections between the concrete and pictorial representations and the abstract maths concept. Children can also create or view a pictorial representation together, enabling discussion and comparisons. The *Power Maths* teaching tools are fantastic for this learning stage, and bar modelling is invaluable for problem solving throughout the primary curriculum.

Abstract

Our ultimate goal is for children to understand abstract mathematical concepts, symbols and notation and of course, some children will reach this stage far more quickly than others. To work with abstract concepts, a child must be comfortable with the meaning of and relationships between concrete, pictorial and abstract models and representations. The C-P-A approach is not linear, and children may need different types of models at different times. However, when a child demonstrates with concrete models and pictorial representations that they have grasped a concept, we can be confident that they are ready to explore or model it with abstract symbols such as numbers and notation.

Use at any time and with any age to support understanding

Variation helps visualisation

Children find it much easier to visualise and grasp concepts if they see them presented in a number of ways, so be prepared to offer and encourage many different representations.

For example, the number six could be represented in various ways:

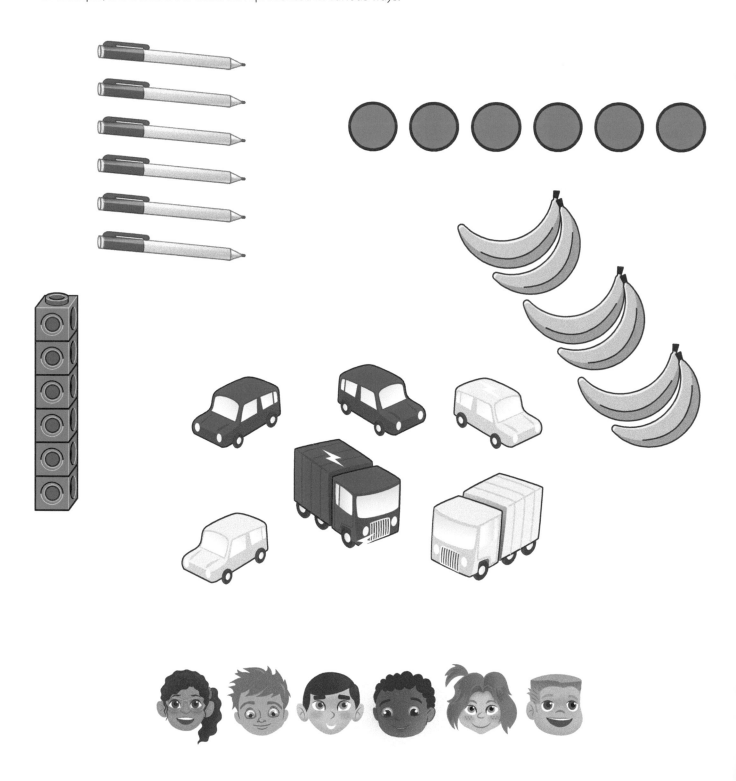

Practical aspects of *Power Maths*

One of the key underlying elements of *Power Maths* is its practical approach, allowing you to make maths real and relevant to your children, no matter their age.

Manipulatives are essential resources for both key stages and *Power Maths* encourages teachers to use these at every opportunity, and to continue the Concrete-Pictorial-Abstract approach right through to Year 6.

The Textbooks and Teacher Guides include lots of opportunities for teaching in a practical way to show children what maths means in real life.

Discover and Share

The **Discover** and **Share** sections of the Textbook give you scope to turn a real-life scenario into a practical and hands-on section of the lesson. Use these sections as inspiration to get active in the classroom. Where appropriate, use the **Discover** contexts as a springboard for your own examples that have particular resonance for your children – and allow them to get their hands dirty trying out the mathematics for themselves.

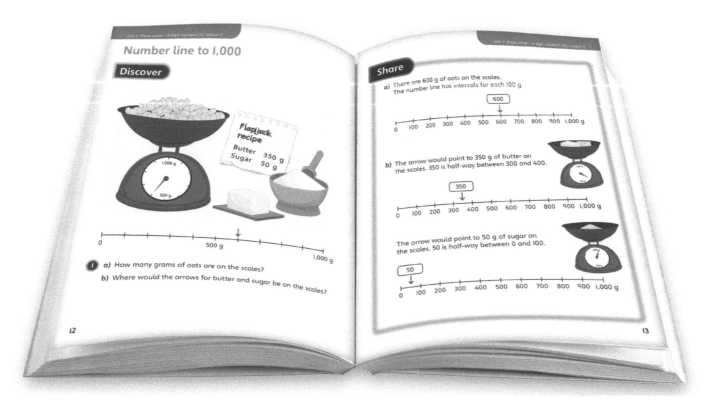

Unit videos

Every term has one unit video which incorporates real-life classroom sequences.

These videos show you how the reasoning behind mathematics can be carried out in a practical manner by showing real children using various concrete and pictorial methods to come to the solution. You can see how using these practical models, such as part-whole and bar models, helps them to find and articulate their answer.

Mastery tips

Mastery Experts give anecdotal advice on where they have used hands-on and real-life elements to inspire their children.

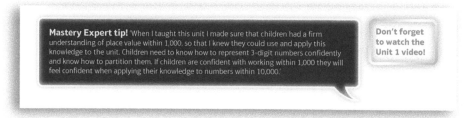

Mastery Expert tip! 'When I taught this unit I made sure that children had a firm understanding of place value within 1,000, so that I knew they could use and apply this knowledge to the unit. Children need to know how to represent 3-digit numbers confidently and know how to partition them. If children are confident with working within 1,000 they will feel confident when applying their knowledge to numbers within 10,000.'

Don't forget to watch the Unit 1 video!

Concrete-Pictorial-Abstract (C-P-A) approach

Each **Share** section uses various methods to explain an answer, helping children to access abstract concepts by using concrete tools, such as counters. Remember, this isn't a linear process, so even children who appear confident using the more abstract method can deepen their knowledge by exploring the concrete representations. Encourage children to use all three methods to really solidify their understanding of a concept.

Pictorial representation – drawing the problem in a logical way that helps children visualise the maths

Concrete representation – using manipulatives to represent the problem. Encourage children to physically use resources to explore the maths.

Abstract representation – using words and calculations to represent the problem.

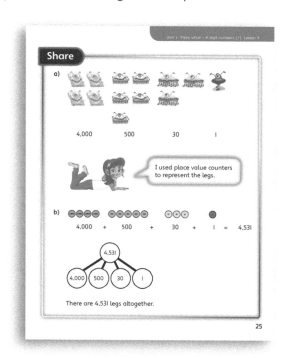

Practical tips

Every lesson suggests how to draw out the practical side of the **Discover** context.

You'll find these in the **Discover** section of the Teacher Guide for each lesson.

PRACTICAL TIPS Begin without a calendar. Ask children to 'guess' the number of days in one year. Talk about any general knowledge that they may have about the number of days in a week, month and year.

Resources

Every lesson lists the practical resources you will need or might want to use. There is also a summary of all of the resources used throughout the term on page 38 to help you be prepared.

RESOURCES

Mandatory: base 10 equipment

Optional: place value counters, specifically numbered number lines

Working with children below age-related expectation

This section offers advice on using *Power Maths* with children who are significantly behind age-related expectation. Teacher judgement will be crucial in terms of where and why children are struggling, and in choosing the right approach. The suggestions can of course be adapted for children with special educational needs, depending on the specific details of those needs.

General approaches to support children who are struggling

Keeping the pace manageable

Remember, you have more teaching days than *Power Maths* lessons so you can cover a lesson over more than one day, and revisit key learning, to ensure all children are ready to move on. You can use the + and – buttons to adjust the time for each unit in the online planning. The NCETM's Ready-to-Progress criteria can be used to help determine what should be highest priority.

Same-day intervention

You could go over the Textbook pages or revisit the previous year's work if necessary (see Addressing gaps). Remember that same-day intervention can be within the lesson, as well as afterwards (see page 27). As children start their independent practice, you can work with those who found the first part of the lesson difficult, checking understanding using manipulatives.

Fluency sessions

Fit in as much practice as you can for number bonds and times-tables, etc., at other times of the day. If you can, plan a short 'maths meeting' for this in the afternoon. You might choose to use a Power Up you haven't used already.

Addressing gaps

Use material from the same topic in the previous year to consolidate or address gaps in learning, e.g. Textbook pages and Strengthen activities. The End of unit check will help gauge children's understanding.

Pre-teaching

Find a 5- to 10-minute slot before the lesson to work with the children you feel would benefit. The afternoon before the lesson can work well, because it gives children time to think in between. Recap previous work on the topic (addressing any gaps you're aware of) and do some fluency practice, targeting number facts etc. that will help children access the learning.

Focusing on the key concepts

If children are a long way behind, it can be helpful to take a step back and think about the key concepts for children to engage with, not just the fine detail of the objective for that year group (e.g. addition with a specific number of columns). Bearing that in mind, how could children advance their understanding of the topic?

Providing extra support within the lesson

Support in the Teacher Guide

First of all, use the Strengthen support in the Teacher Guide for guided and independent work in each lesson, and share this with Teaching Assistants, where relevant. As you read through the lesson content and corresponding Teacher Guide pages before the lesson, ask yourself what key idea or nugget of understanding is at the heart of the lesson. If children are struggling, this should help you decide what's essential for all children before they move on.

Annotating pages

You can annotate questions to provide extra scaffolding or hints if you need to, but aim to build up children's ability to access questions independently wherever you can. Children tend to get used to the style of the *Power Maths* questions over time.

Quick recap as lesson starter

The Quick recap for each lesson in the Teacher Guide is an alternative starter activity to the Power Up. You might choose to use this with some or all children if you feel they will need support accessing the main lesson.

Consolidation questions

If you think some children would benefit from additional questions at the same level before moving on, write one or two similar questions on the board. (This shouldn't be at the expense of reasoning and problem-solving opportunities: take longer over the lesson if you need to.)

Hard copy Textbooks

The Textbooks help children focus in more easily on the mathematical representations, read the text more comfortably, and revisit work from a previous lesson that you are building on, as well as giving children ownership of their learning journey. In main lessons, it can work well to use the e-Textbook for **Discover** and give out the books when discussing the methods in the **Share** section.

Reading support

It's important that all children are exposed to problem solving and reasoning questions, which often involve reading. For whole-class work you can read questions together. For independent practice you could consider annotating pages to help children see what the question is asking, and stem sentences to help structure their answer. A general focus on specific mathematical language and vocabulary will help children access the questions. You could consider pairing weaker readers with stronger readers, or read questions as a group if those who need support are on the same table.

Providing extra depth and challenge with *Power Maths*

Just as prescribed in the National Curriculum, the goal of *Power Maths* is never to accelerate through a topic but rather to gain a clear, deep and broad understanding. Here are some suggestions to help ensure all children are appropriately challenged as you work with the resources.

Overall approaches

First of all, remember that the materials are designed to help you keep the class together, allowing all children to master a concept while those who grasp it quickly have time to explore it in more depth. Use the Deepen support in the Teacher Guide (see below) to challenge children who work through the questions quickly. Here are some questions and ideas to encourage breadth and depth during specific parts of the lesson, or at any time (where no part of the lesson sequence is specified):

- **Discover**: 'Can you demonstrate your solution another way?'

- **Share**: Make sure every child is encouraged to give answers and engage with the discussion, not just the most confident.

- **Think together**: 'Can you model your answers using concrete materials? Can you explain your solution to a partner?'

- Practice: Allow all children to work through the full set of questions, so that they benefit from the logical sequence.

- **Reflect**: 'Is there another way of working out the answer? And another way?'
  'Have you found all the solutions?'
  'Is that always true?'
  'What's different between this question and that question? And what's the same?'

Note that the **Challenge** questions are designed so that all children can access and attempt them, if they have worked through the steps leading up to them. There may be some children in a given lesson who don't manage to do the **Challenge**, but it is not supposed to be a distinct task for a subset of the class. When you look through the lesson materials before teaching, think about what each question is specifically asking, and compare this with the key learning point for the lesson. This will help you decide which questions you feel it's essential for all children to answer, before moving on. You can at least aim for all children to try the **Challenge**!

Deepen activities and support

The Teacher Guide provides valuable support for each stage of the lesson. This includes Deepen tips for the guided and independent practice sections, which will help you provide extra stretch and challenge within your lesson, without having to organise additional tasks. If you have a Teaching Assistant, they can also make use of this advice. There are also suggestions for the lesson as a whole in the 'Going Deeper' section on the first page of the Teacher Guide section for that lesson. Every class is different, so you can always go a bit further in the direction indicated, if appropriate, and build on the suggestions given.

There is a Deepen activity for each unit. These are designed to follow on from the End of unit check, stretching children who have a firm understanding of the key learning from the unit. Children can work on them independently, which makes it easier for the teacher to facilitate the Strengthen activity for children who need extra support. Deepen activities could also be introduced earlier in the unit if the necessary work has been covered. The Deepen activities are on *ActiveLearn* on the Planning page for each unit, and also on the Resources page).

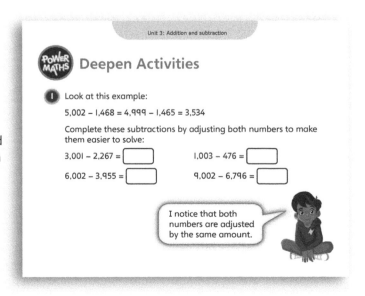

Unit 3: Addition and subtraction

Deepen Activities

1. Look at this example:

$5,002 - 1,468 = 4,999 - 1,465 = 3,534$

Complete these subtractions by adjusting both numbers to make them easier to solve:

$3,001 - 2,267 = \boxed{}$ $\qquad$ $1,003 - 476 = \boxed{}$

$6,002 - 3,955 = \boxed{}$ $\qquad$ $9,002 - 6,796 = \boxed{}$

I notice that both numbers are adjusted by the same amount.

Using the questions flexibly to provide extra challenge

Sometimes you may want to write an extra question on the board or provide this on paper. You can usually do this by tweaking the lesson materials. The questions are designed to form a carefully structured sequence that builds understanding step by step, but, with careful thought about the purpose of each question, you can use the materials flexibly where you need to. Sometimes you might feel that children would benefit from another similar question for consolidation before moving on to the next one, or you might feel that they would benefit from a harder example in the same style. It should be quick and easy to generate 'more of the same' type questions where this is the case.

For this example (from Unit 3, Lesson 6), you could ask children to make up their own question(s) for a partner to solve. If you blot out more than four digits, does that make it easier or harder? Can children still devise questions with one exchange? (For any of the examples on this page you could ask early finishers to create their own question for a partner. 'Guess my number' questions are also good for this, e.g. Practice Book 4A page 20.).

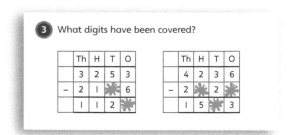

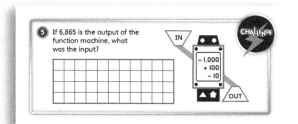

When you see a question like this one (from Unit 1, Lesson 7), it's easy to make extra examples to do afterwards if you need them. You could choose any 4-digit number, or choose tricky examples designed to cross 1000s (e.g. 1967 or 9909).

Here's an example (from Unit 3, Lesson 16) where some of the information in the picture is used for questions in the lesson, but not all. Clearly there are extra questions you could ask using the same information. For example, what's the biggest and smallest difference between two camps? Or, how much climbing is it to go from Camp 4 down to Camp 1 and then back up to Camp 3?

Besides creating additional questions, you should be able to find a question in the lesson that you can adapt into a game or open-ended investigation, if this helps to keep everyone engaged. It could simply be that, instead of answering 5 × 5 etc. on the page, they could build a robot with 5 lots of 5 cubes. Many lessons introduce a game anyway (e.g. Textbook 4A page 71).

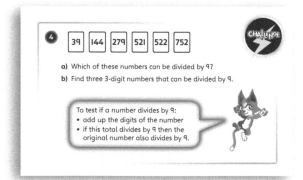

With a question like this (from Unit 5, Lesson 4), you could introduce a game where one child writes a list of numbers where only one is divisible by 9, and their partner has to work out which number it is.

See the bullets above for some general ideas that will help with 'opening out' questions in the books, e.g. 'can you find all the solutions?' type questions.

Other suggestions

Another way of stretching children is through mixed ability pairs, or via other opportunities for children to explain their understanding in their own way. This is a good way of encouraging children to go deeper into the learning, rather than, for instance, tackling questions that are computationally more challenging but conceptually equivalent in level.

Using *Power Maths* with mixed age classes

Overall approaches

There are many variables between schools that would make it inadvisable to recommend a one-size-fits-all approach to mixed age teaching with *Power Maths*. These include how year groups are merged, availability of Teaching Assistants, experience and preference of teaching staff, range in pupil attainment across years, classroom space and layout, level of flexibility around timetables, and overall organisational structure (whether the school is part of a trust).

Some schools will find it best to timetable separate maths lessons for the different year groups. Others will aim to teach the class together as much as possible using the mixed age planning support on *ActiveLearn* (see the lesson exemplars for ways of organising lessons with strong/medium/weak correlation between year groups). There will also be ways of adapting these general approaches. For example, offset lessons where Year A start their lesson with the teacher, while Year B work independently on the practice from the previous lesson, and then start the next lesson with the teacher while Year A work independently; or teachers may choose to base their provision around the lesson from one year group and tweak the content up/down for the other group.

Key strategies for mixed age teaching

The mixed age teaching webinar on *ActiveLearn* provides advice on all aspects of mixed age teaching, including more detail on the ideas below.

Developing independence over time
Investing time in building up children's independence will pay off in the medium term.

Clear rationale
If someone asked, 'Why did you teach both Unit 3 and 4 in the same lesson/separate lessons?', what would your answer be?

Designing a lesson
1. Identify the core learning for each group
2. Identify any number skills necessary to access the core
3. Consider the flow of concepts and how one core leads to the other

Challenging all children
The questions are designed to build understanding step by step, but with careful thought about the purpose of each question you can tweak them to increase the challenge.

Multiple years combined
With more than two years together, teachers will inevitably need to use the resources flexibly if delivering a single lesson.

Enjoy the positives!

Comparison deepens understanding and there will be lots of opportunities for children, as well as misconceptions to explore. There is also in-built pre-teaching and the chance to build up a concept from its foundations. For teachers there is double the material to draw on! Mixed age teachers require a strong understanding of the progression of ideas across year groups, which is highly valuable for all teachers. Also, it is necessary to engage deeply with the lesson to see how to use the materials flexibly – this is recommended for all teachers and will help you bring your lesson to life!

List of practical resources

Year 4A Mandatory resources

Resource	Lesson
100 squares	**Unit 5** Lesson 6
base 10 equipment	**Unit 1** Lessons 1, 3, 7, 8 **Unit 3** Lessons 1, 2, 3, 4, 5, 6, 7, 8 **Unit 5** Lessons 4, 7, 8, 9, 10, 11
counters	**Unit 1** Lessons 5, 8 **Unit 4** Lessons 2, 4 **Unit 5** Lessons 1, 2, 3, 4, 5, 7, 8, 10, 11, 12
counters (3 different colours)	**Unit 5** Lesson 6
counters (small, preferably 16 mm)	**Unit 4** Lesson 1
cubes	**Unit 5** Lessons 1, 9, 12
dice	**Unit 1** Lessons 1, 4 **Unit 3** Lessons 5, 7, 8, 10
number cards (or sticky notes)	**Unit 5** Lesson 6
number lines	**Unit 2** Lessons 1, 2, 5, 7, 8
number lines (blank)	**Unit 2** Lesson 6
number lines (printed, from 1,000 to 2,000 and from 3,000 to 4,000)	**Unit 3** Lesson 11
objects to measure with (non-standard, e.g. flat coloured squares or triangles, playing cards, coins)	**Unit 4** Lesson 1
paper (square dotted)	**Unit 4** Lesson 4
paper (squared)	**Unit 4** Lessons 2, 3, 4
paper squares	**Unit 4** Lesson 2
paper squares (1 cm)	**Unit 4** Lesson 4
place value counters	**Unit 1** Lessons 7, 8 **Unit 3** Lessons 1, 2, 3, 4, 5, 6, 7, 8
place value equipment	**Unit 1** Lessons 1, 4, 5, 6 **Unit 3** Lesson 10
place value grids	**Unit 2** Lessons 4, 8
plastic shapes (flat)	**Unit 4** Lesson 3

Year 4A Optional resources

Resource	Lesson
100 square	**Unit 5** Lesson 1
base 10 equipment	**Unit 1** Lesson 7 **Unit 2** Lessons 4, 8 **Unit 5** Lessons 2, 3, 5
chalk	**Unit 4** Lessons 3, 4
counters	**Unit 2** Lesson 8 **Unit 4** Lesson 3 **Unit 5** Lesson 9
dice	**Unit 2** Lesson 6 **Unit 5** Lesson 2
digit cards	**Unit 1** Lesson 4 **Unit 2** Lesson 7
flashcards (4 times-table)	**Unit 5** Lesson 12
geoboards	**Unit 4** Lesson 2
grid (large, with patterned squares or pens to colour in the squares)	**Unit 4** Lesson 5
manipulatives (counters, small squares)	**Unit 4** Lesson 5
masking tape	**Unit 4** Lessons 3, 4
measuring equipment (with number line scales)	**Unit 1** Lesson 2
number lines	**Unit 1** Lesson 7 **Unit 2** Lesson 3 **Unit 2** Lesson 4
number lines (specifically numbered)	**Unit 1** Lesson 3 **Unit 2** Lesson 6
paper strips (to make bar models)	**Unit 3** Lessons 5, 13
pegs	**Unit 2** Lesson 2
place value counters	**Unit 1** Lesson 3 **Unit 2** Lesson 4
place value equipment	**Unit 3** Lesson 9
place value grids	**Unit 1** Lesson 7
rectangles (cut out from coloured card)	**Unit 4** Lesson 3
scales	**Unit 2** Lesson 1
square tiles	**Unit 4** Lesson 4
squares (cut out from coloured card)	**Unit 4** Lesson 2
squares and rectangles (to measure with, e.g. book covers, newspaper pages, paper or card)	**Unit 4** Lesson 1
sticky notes	**Unit 2** Lessons 2, 3, 7 **Unit 3** Lesson 6 **Unit 4** Lesson 3
string	**Unit 3** Lesson 6

Getting started with *Power Maths*

As you prepare to put *Power Maths* into action, you might find the tips and advice below helpful.

STEP 1: Train up!

A practical, up-front full day professional development course will give you and your team a brilliant head-start as you begin your *Power Maths* journey. You will learn more about the ethos, how it works and why.

STEP 2: Check out the progression

Take a look at the yearly and termly overviews. Next take a look at the unit overview for the unit you are about to teach in your Teacher Guide, remembering that you can match your lessons and pacing to match your class.

STEP 3: Explore the context

Take a little time to look at the context for this unit: what are the implications for the unit ahead? (Think about key language, common misunderstandings and intervention strategies, for example.) If you have the online subscription, don't forget to watch the corresponding unit video.

STEP 4: Prepare for your first lesson

Familiarise yourself with the objectives, essential questions to ask and the resources you will need. The Teacher Guide offers tips, ideas and guidance on individual lessons to help you anticipate children's misconceptions and challenge those who are ready to think more deeply.

STEP 5: Teach and reflect

Deliver your lesson — and enjoy!

Afterwards, reflect on how it went… Did you cover all five stages?
Does the lesson need more time? How could you improve it?
What percentage of your class do you think mastered the concept?
How can you help those that didn't?

Unit 1
Place value – 4-digit numbers ①

Mastery Expert tip! 'When I taught this unit I made sure that children had a firm understanding of place value within 1,000, so that I knew they could use and apply this knowledge to the unit. Children need to know how to represent 3-digit numbers confidently and know how to partition them. If children are confident with working within 1,000 they will feel confident when applying their knowledge to numbers within 10,000.'

Don't forget to watch the Unit 1 video!

WHY THIS UNIT IS IMPORTANT

This is a pivotal unit because a solid understanding of place value using 4-digit numbers is fundamental to success in other areas of learning, particularly the four operations of addition, subtraction, multiplication and division.

Equally, children who struggle to represent and count with 4-digit numbers may struggle when they are asked to apply these skills in future learning. It is important to ensure that enough time is spent on this unit of work so that children have a solid understanding of place value both now and for the future.

WHERE THIS UNIT FITS

→ **Unit 1: Place value – 4-digit numbers (1)**
→ Unit 2: Place value – 4-digit numbers (2)

This unit builds on previous learning in Year 3 about place value within 1,000. This previous learning introduced children to the concept of counting in 10s, comparing numbers, ordering numbers and using a number line to 1,000. Children will continue to use these previously learnt skills and apply them when working with 4-digit numbers. Before they start this unit, it is expected that children:

• have a solid understanding of place value within 1,000 from Year 3
• understand how to count in 10s and 100s
• can order and compare numbers to 1,000.

ASSESSING MASTERY

Children who have mastered this unit will be able to represent 4-digit numbers using a variety of concrete apparatus. They will know that a number is made up of 1,000s, 100s, 10s and 1s and will be able to represent numbers in multiple ways. They will use this learning to find 1, 10, 100, 1,000 more or less than a given number.

COMMON MISCONCEPTIONS	STRENGTHENING UNDERSTANDING	GOING DEEPER
Children may think that 3 tens + 2 thousands + 5 hundreds + 7 ones is 3,257 as opposed to 2,537.	Use base 10 equipment and place value counters to secure understanding of 4-digit numbers.	Ask children to partition numbers in different ways. For example, 1,223 is $1 \times 1,000$, 1×100, 2×10 and 3×1.
When finding 1,000 more or less than a number and comparing numbers, children may compare using the incorrect column.	Allow children to physically build the number using concrete apparatus so that they can physically see the digits. In addition to this, present the number in a place value grid so that children can see which place value column is the important one.	Children can order and compare a set of numbers to deepen learning.

Unit I: Place value – 4-digit numbers

UNIT STARTER PAGES

Introduce this unit of work using teacher-led discussion. Allow children time to discuss questions in pairs or small groups and share ideas as a whole class. Children should be encouraged to use concrete apparatus so that they can see the numbers they will be working with.

STRUCTURES AND REPRESENTATIONS

Number line: The number line will allow children to see which numbers a number sits between. It also allows children to see how increments can be used to view numbers.

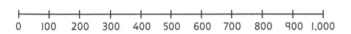

| 0 | 100 | 200 | 300 | 400 | 500 | 600 | 700 | 800 | 900 | 1,000 |

Place value equipment, including using base 10, and place value grid with place value counters: This model will help children organise 4-digit numbers into 1,000s, 100s, 10s and 1s, with both concrete representations and abstract numbers.

Part-whole models: This model will help children partition a 4-digit number into thousands, hundreds, tens and units.

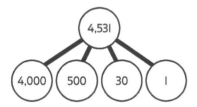

KEY LANGUAGE

→ There is some key language that children will need to know as part of the learning in this unit:

→ tens (10s), hundreds (100s), thousands (1,000s)

→ counting, represent

→ partition, recombine

→ numerals

→ parts, whole

→ place value

→ number line

→ multiples

→ digit

PUPIL TEXTBOOK 4A PAGE 6

PUPIL TEXTBOOK 4A PAGE 7

Represent and partition numbers to 1,000

Learning focus

In this lesson, children rehearse the place value of 3-digit numbers and explore different ways to partition them. They make the numbers using different equipment and representations, and find the value of each digit.

Before you teach

- Can children count in ones from any 3-digit number, including crossing a 100? For example, 298, 299, 300, 301, and so on?
- Have they had opportunities to practise building 2-digit numbers using place value equipment?

NATIONAL CURRICULUM LINKS

Year 4 Number – number and place value

Recognise the place value of each digit in a 4-digit number (1,000s, 100s, 10s, and 1s).

ASSESSING MASTERY

Children can represent 3-digit numbers in a variety of ways and can use this to find the value of each digit.

COMMON MISCONCEPTIONS

Children may not understand how to find the value of each digit and may partition 794 into 7, 9 and 4, rather than 700, 90 and 4. They may also struggle to represent and explain the place value of digits in numbers that have 0 as a place holder. Ask:

- *What is the value of each digit in the number 109?*
- *What is the same and what is different in these numbers: 230, 233, 203?*

STRENGTHENING UNDERSTANDING

Encourage children to use place value equipment alongside any discussion of place value, to make each number that is mentioned and to describe what this tells them about the value of each digit.

GOING DEEPER

Challenge children to explore a range of 3-digit numbers and to find interesting patterns. For example, numbers with repeating digits (such as 333); numbers with consecutive digits (such as 345); numbers with descending digits (such as 543); numbers with all even digits (such as 246); numbers with all odd digits (such as 975).

KEY LANGUAGE

In lesson: place value, hundreds (100s), tens (10s), ones (1s)

STRUCTURES AND REPRESENTATIONS

Part-whole model, place value grid

RESOURCES

Mandatory: base 10 equipment, place value equipment, dice

 In the eTextbook of this lesson, you will find interactive links to a selection of teaching tools.

Quick recap

Ask children to take turns to roll three dice and to use the digits to make a 3-digit number. Repeat several times. As a class, practise saying the numbers together. Ask questions to help compare the numbers, for example: *Who made a number with 3 hundreds? Who made a number greater than 400?*

Discover

WAYS OF WORKING Pair work

ASK

- Question ❶: *Do you know how many days there are in a week, a month or even one year?*
- Question ❶ a): *What do you notice about the number 365?*

IN FOCUS This shows a discussion about some known 3-digit numbers and is an example of recognising numbers that are important in general knowledge and practical situations.

PRACTICAL TIPS Begin without a calendar. Ask children to 'guess' the number of days in one year. Talk about any general knowledge that they may have about the number of days in a week, month and year.

ANSWERS

Question ❶ a): The base 10 equipment should show
 3 hundreds, 6 tens and 5 ones.

Question ❶ b): A part-whole model should show
 366 = 300 + 60 + 6.
 Place value counters should show
 3 hundreds, 6 tens and 6 ones.

Represent and partition numbers to 1,000

Discover

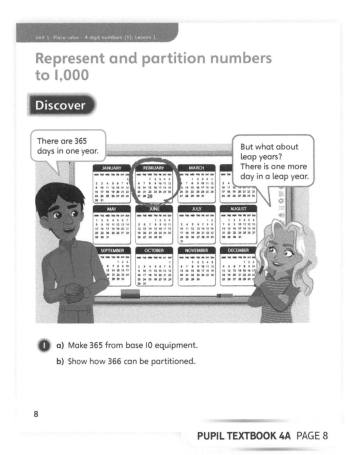

❶ a) Make 365 from base 10 equipment.
 b) Show how 366 can be partitioned.

8

PUPIL TEXTBOOK 4A PAGE 8

Share

WAYS OF WORKING Whole class teacher led

ASK

- Question ❶ a): *What does each digit represent?*
- Question ❶ b): *Why do you think we have multiple ways to show 3-digit numbers?*

IN FOCUS The focus is on familiarising children with 3-digit numbers and several different ways of representing them. Question ❶ a) rehearses the place value of 3-digit numbers, whilst question ❶ b) explores variations in place value representations.

Share

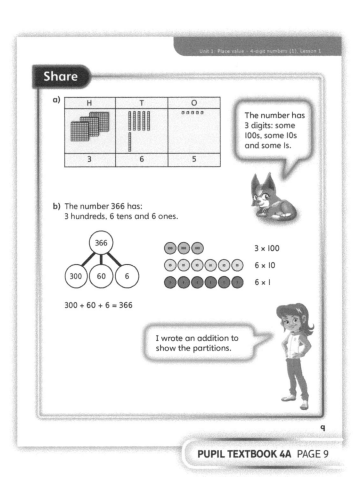

PUPIL TEXTBOOK 4A PAGE 9

Think together

Whole class teacher led (I do, We do, You do)

ASK

- Question ❶: *How do the different parts of your representation show the place value of 234?*
- Question ❷: *What is the whole number? What are the parts?*
- Question ❸: *What is the same and what is different in these diagrams?*

IN FOCUS Question ❶ shows a basic representation of a 3-digit number using place value equipment.
Question ❷ a) uses the part-whole model to partition numbers and show the value of each digit, whereas question ❷ b) shows partitioning as addition.
Question ❸ requires children to recognise the use of zero as a place holder in 3-digit numbers.

STRENGTHEN Use equipment to explore numbers initially. Teach children how to draw simple pictures of 3-digit equipment to develop more independence from the resource.

DEEPEN Ask children to explore the use of part-whole models for numbers with '0' as a place holder.

ASSESSMENT CHECKPOINT Question ❷ assesses whether children can accurately complete addition sentences to show partitioning.

ANSWERS

Question ❶: 200 + 30 + 4

Question ❷ a): 684
700 + 90 + 4

Question ❷ b): **684** = 600 + 80 + 4
794 = **700 + 90 + 4**

Question ❷ c): The additions use the same numbers that were partitioned in the part-whole models.

Question ❸ a): A: 244, B: 240, C: 204, D: 420

Question ❸ b): C: 204

Question ❸ c): Children's sketches should show the hundreds as squares, the tens as sticks and the ones as cubes.
305: 3 hundreds and 5 ones
350: 3 hundreds and 5 tens
353: 3 hundreds, 5 tens and 3 ones
503: 5 hundreds and 3 ones

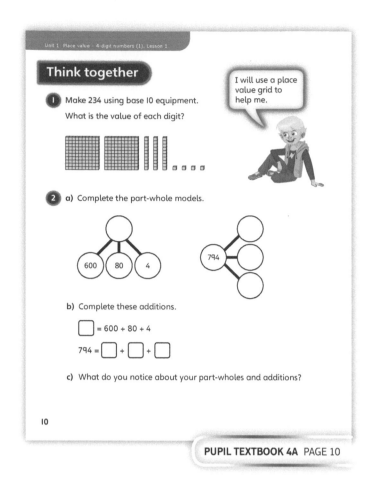

PUPIL TEXTBOOK 4A PAGE 10

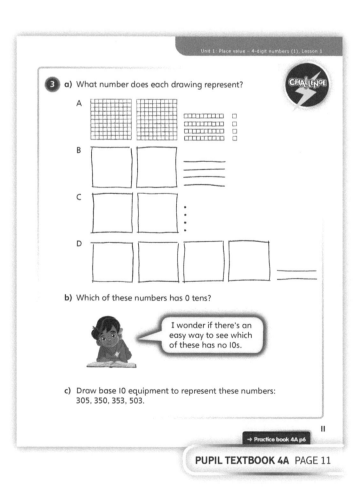

PUPIL TEXTBOOK 4A PAGE 11

Practice

WAYS OF WORKING Independent thinking

IN FOCUS

Question **1** shows basic place value representations of 3-digit numbers, whilst children draw their own simple representations in question **2**. In question **3** children recognise the value of each digit in 3-digit numbers and they partition into 100s, 10s and 1s in question **4**.

3-digit numbers are partitioned and also recombined in question **5**. To approach this question, children need to understand the link between addition and partitioning. In question **7** children explore variations in 3-digit numbers shown on a place value grid.

STRENGTHEN Work with children to help them understand the link between 300, 3 hundreds and the '3' digit in the 100s place.

DEEPEN Explore variations on question **7**. Ask: *How many different 3-digit numbers can you make using seven counters? Can you convince me that you have found all the options?*

ASSESSMENT CHECKPOINT Use question **5** to assess whether children can partition and recombine 3-digit numbers into 100s, 10s and 1s.

ANSWERS Answers for the **Practice** part of the lesson can be found in the *Power Maths* online subscription.

Reflect

WAYS OF WORKING Independent thinking

IN FOCUS The **Reflect** part of the lesson gives children the chance to explore variation in 3-digit numbers that meet specific place value criteria.

ASSESSMENT CHECKPOINT Assess whether children can explain the pattern of numbers that meet the conditions given.

ANSWERS Answers for the **Reflect** part of the lesson can be found in the *Power Maths* online subscription.

After the lesson ⏸

- Can children show different 3-digit numbers using a variety of representations and equipment?
- Can children name 3-digit numbers that you have made using different representations?

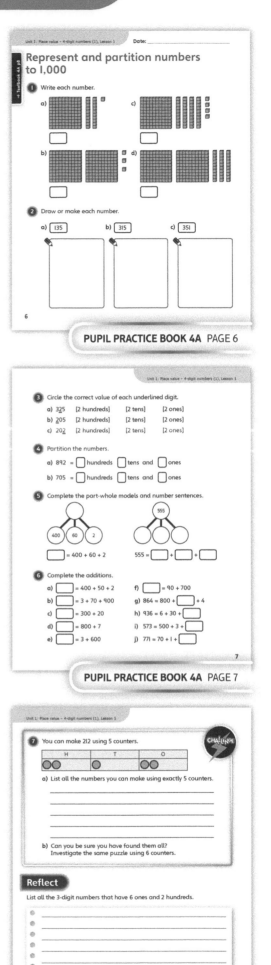

PUPIL PRACTICE BOOK 4A PAGE 6

PUPIL PRACTICE BOOK 4A PAGE 7

PUPIL PRACTICE BOOK 4A PAGE 8

Number line to 1,000

Learning focus

In this lesson, children rehearse finding the position of 3-digit numbers on a variety of number lines, including some real-life contexts.

Before you teach

- Ask children to place 2-digit numbers on a variety of different number lines; can they draw a 0 to 100 number line that counts in 10s?
- Can they draw a number line from 40 to 50, and indicate the position of 47 on it?

NATIONAL CURRICULUM LINKS

Year 4 Number – number and place value

Recognise the place value of each digit in a four-digit number (1,000s, 100s, 10s, and 1s).

ASSESSING MASTERY

Children can label intervals and recognise given numbers on number lines. They can also place 3-digit numbers accurately on number lines.

COMMON MISCONCEPTIONS

Children may find some number line conventions confusing, for example where there are variations in the starting number, the intervals or the length. Ask:

- *What number starts this number line?*
- *What jumps can you make on this number line?*
- *Does this number line count in 1s, 10s or 100s?*

STRENGTHENING UNDERSTANDING

Encourage children to count up and down along the number line whilst you point. Build confidence by working specifically on number lines that count in 100s, count in 10s and count in 1s.

GOING DEEPER

Ask children to explore number lines that are used in measuring equipment for length, mass or volume. Provide real-life examples or ask children to find some of their own.

KEY LANGUAGE

In lesson: interval, half-way

STRUCTURES AND REPRESENTATIONS

Number line

RESOURCES

Optional: measuring equipment with number line scales

 In the eTextbook of this lesson, you will find interactive links to a selection of teaching tools.

Quick recap

Give children a sheet with some written information about animals:

Lion: may weigh between 160 kg and 250 kg

Tiger: may weigh between 90 kg and 300 kg

Ask children to read the number ranges aloud and then show them on a number line. Together, discuss possible 3-digit numbers that fall within the ranges.

Discover

WAYS OF WORKING Pair work

ASK

- *What do you notice about the measuring scale?*
- *Where does the number line start? What does it count up in?*

IN FOCUS In this activity, children are counting in steps of 100 on a number line from 0 to 1,000 g in the context of using a measuring scale when cooking.

PRACTICAL TIPS Use measuring scales with a dial like the one in the question to measure real ingredients. Encourage children to watch as the dial moves when more mass is added on to the scale.

ANSWERS

Question ❶ a): 600g

Question ❶ b): Butter: half-way between 3rd and 4th markers.
Sugar: half-way between 0 and next marker.

Number line to 1,000

Discover

a) How many grams of oats are on the scales?

b) Where would the arrows for butter and sugar be on the scales?

12

PUPIL TEXTBOOK 4A PAGE 12

Share

WAYS OF WORKING Whole class teacher led

ASK

- Question ❶ a): *Can you count up and down on this number line? Why do you think it counts up in steps of 100?*
- Question ❶ b): *Can you see how to find any other numbers between the 100s?*

IN FOCUS The focus is on recognising intervals of 100 on a 0 to 1,000 number line, where not all the intervals are labelled. The numbers given require children to identify numbers that come half-way between the given intervals.

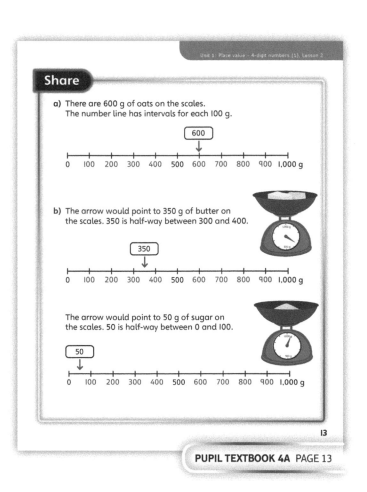

Share

a) There are 600 g of oats on the scales. The number line has intervals for each 100 g.

b) The arrow would point to 350 g of butter on the scales. 350 is half-way between 300 and 400.

The arrow would point to 50 g of sugar on the scales. 50 is half-way between 0 and 100.

13

PUPIL TEXTBOOK 4A PAGE 13

Think together

WAYS OF WORKING Whole class teacher led (I do, We do, You do)

ASK

- Question ❶: *What is the same and what is different about each number line?*
- Question ❷: *How do you know whether to count in 1s, 10s or 100s?*
- Question ❸: *What is important when making an estimate?*

IN FOCUS In question ❶, children are identifying specific intervals and using this to find given numbers.

Question ❷ requires children to estimate the position of a given number on number lines with different scales and intervals. In question ❸ children estimate on open number lines.

STRENGTHEN Ensure that children spend time drawing number lines that count first in 100s, then in 10s and then in 1s.

DEEPEN Ask children to investigate the different scales that are used on a variety of measuring equipment.

ASSESSMENT CHECKPOINT For question ❷, assess whether children are able to explain the reasoning behind their choice for the position of the number on each of the number lines.

ANSWERS

Question ❶ a): 100
600
800

Question ❶ b): 272
274
278

Question ❷ a): Half-way between 500 and previous marker (400).

Question ❷ b): Just before the 450 marker.

Question ❷ c): On marker to the left of 450.

Question ❸ a): Approximately 750 ml [740–760 acceptable].

Question ❸ b): Approximately 270 [260–280 acceptable].

Question ❸ c): Approximately 850 [820–880 acceptable].
Approximately 550 [520–580 acceptable].

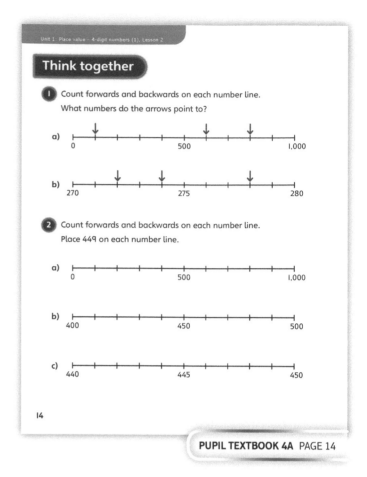

PUPIL TEXTBOOK 4A PAGE 14

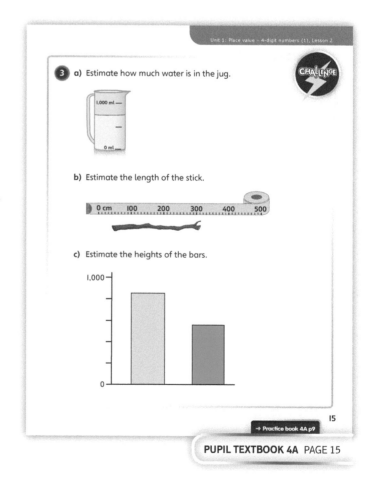

PUPIL TEXTBOOK 4A PAGE 15

Practice

WAYS OF WORKING Independent thinking

IN FOCUS Question ❶ requires children to recognise the intervals that have been used on a variety of number lines. In question ❷ children are locating numbers on given lines and in question ❸ they need to recognise which numbers are indicated by their position. For question ❹ children are reading number lines as measuring scales.

STRENGTHEN Let children practise moving an arrow or peg up and down on a large 0 to 1,000 number line. Discuss good estimates for the position of given numbers.

DEEPEN In question ❻ children draw their own number lines. Explore drawing very big lines, for example using the line markings on the playground. Challenge children to locate intervals for 100s, 10s and 1s on a line of this size.

ASSESSMENT CHECKPOINT Question ❸ assesses whether children can recognise numbers by their position on different number lines. Can they make reasoned justifications for their answers in this question?

ANSWERS Answers for the **Practice** part of the lesson can be found in the *Power Maths* online subscription.

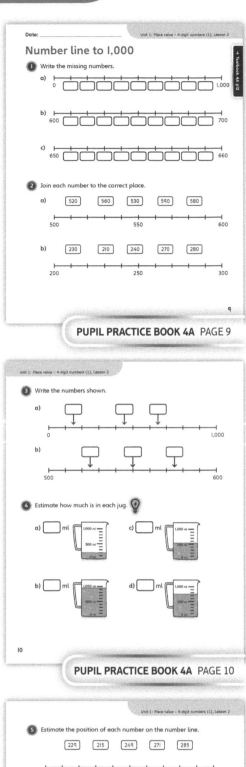

PUPIL PRACTICE BOOK 4A PAGE 9

Reflect

WAYS OF WORKING Independent thinking

IN FOCUS The **Reflect** part of the lesson is an open prompt to allow children to talk about the different positions of numbers on number lines. This will depend on variations in the start and end numbers and also the intervals given.

ASSESSMENT CHECKPOINT Assess whether children can justify their decisions and recognise the ambiguity in the question.

ANSWERS Answers for the **Reflect** part of the lesson can be found in the *Power Maths* online subscription.

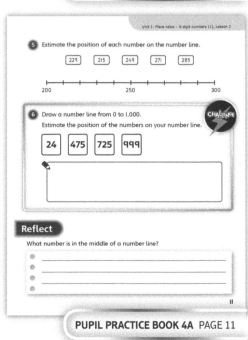

PUPIL PRACTICE BOOK 4A PAGE 11

After the lesson ⏸

- Draw a class number line together on the board. Can children recognise and place given 3-digit numbers on it?

Multiples of 1,000

Learning focus

In this lesson, children will count in 1,000s from 0 to 10,000, forwards and backwards and recognise multiples of 1,000 in different representations.

Before you teach

- Can children count in 10s?
- Can children count in 100s?
- Can children use a comma correctly when writing numbers in 1,000s?

NATIONAL CURRICULUM LINKS

Year 4 Number – number and place value

Count in multiples of 6, 7, 9, 25 and 1,000.

ASSESSING MASTERY

Children can count in 1,000s from 0 to 10,000, forwards and backwards. Children should recognise what multiples of 1,000 look like and be able to write numbers in words and numerals.

COMMON MISCONCEPTIONS

Children may struggle to start counting mid-sequence (for example, starting at 5,000 rather than at 0). Encourage children to point to a representation of each 1,000 as they say it. Ask:

- *Can you point to the number as you count?*
- *Can you count backwards in 1,000s?*

STRENGTHENING UNDERSTANDING

To strengthen understanding, ask children to count in 1,000s using base 10 equipment. Use the base 10 equipment to support children's ability to visualise 1,000. Discuss how the 1,000 block is made up of 10 hundreds.

GOING DEEPER

Ask children to count both forwards and backwards in 1,000s. Ask them to count in 1,000s above 10,000. Ask: *What number comes next? How do you know?*

KEY LANGUAGE

In lesson: thousands (1,000s), represent, number sequences

Other language used by the teacher: numeral, hundreds (100s), tens (10s), number line

STRUCTURES AND REPRESENTATIONS

Place value grids,, number lines, number tracks

RESOURCES

Mandatory: base 10 equipment

Optional: place value counters, specifically numbered number lines

 In the eTextbook of this lesson, you will find interactive links to a selection of teaching tools.

Quick recap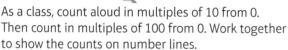

As a class, count aloud in multiples of 10 from 0. Then count in multiples of 100 from 0. Work together to show the counts on number lines.

Discover

WAYS OF WORKING Pair work

ASK

- Question ❶ a): *How many boxes of lemon sweets are there?*
- Question ❶ a): *How many sweets are there in each box?*
- Question ❶ a): *Should we count in 10s, 100s or 1,000s? Why?*

IN FOCUS Questions ❶ a) and b) stress the importance of counting from 0 in 1,000s to see how many sweets there are. It is important that children count in 1,000s at this stage, rather than just saying the answer, so encourage them to point to the boxes of sweets as they count.

PRACTICAL TIPS Introduce children to the 1,000 block in base 10 equipment and encourage them to use these blocks to represent the boxes of sweets as they count.

ANSWERS

Question ❶ a): There are 4,000 lemon sweets.
　　　　　　　There are 6,000 strawberry sweets.

Question ❶ b): There are 10,000 sweets altogether:
　　　　　　　4,000 + 6,000 = 10,000

Multiples of 1,000

Discover

❶ a) Count the lemon sweets on the forklift pallet.
　　　Count the strawberry sweets on the forklift pallet.

b) How many sweets are there altogether?

16

PUPIL TEXTBOOK 4A PAGE 16

Share

WAYS OF WORKING Whole class teacher led

ASK

- Question ❶ a): *How many sweets does each box represent? How can you use this information to find the total number of sweets?*
- Question ❶ b): *Complete the count. Did your work on part a) help you?*

IN FOCUS In question ❶ b) children may see that they can start the count from 4,000 and count on to 10,000, based on the answer to part a). This is a useful discussion point. Ask: *Do we always need to start the count at 0?*

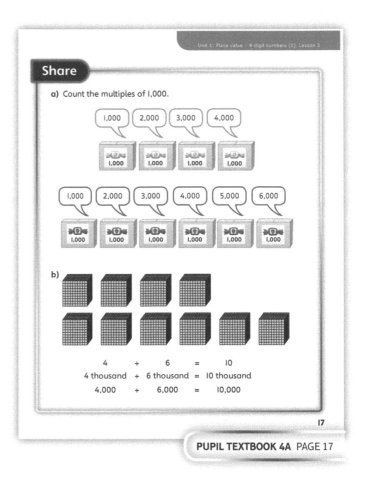

Share

a) Count the multiples of 1,000.

1,000　2,000　3,000　4,000

5,000　6,000 ... (boxes)

b)

4　+　6　=　10
4 thousand　+　6 thousand　=　10 thousand
4,000　+　6,000　=　10,000

17

PUPIL TEXTBOOK 4A PAGE 17

Think together

WAYS OF WORKING Whole class teacher led (I do, We do, You do)

ASK
- Question **1**: *How many boxes of strawberry and lemon sweets are there? How many sweets are in each box?*
- Question **2**: *How can the numbers shown help us?*
- Question **3**: *How many 10s are in 100? How might this help you?*

IN FOCUS Question **2** encourages children to apply their knowledge from the beginning of the lesson to working out the missing numbers in a sequence or number track. Question **3** introduces 1,000 as 10 × 100 and should strengthen children's understanding of the base 10 system.

STRENGTHEN Encourage children to represent numbers using base 10 equipment and to point to the 1,000s as they count.

DEEPEN Question **3** could lead to other questions, such as: *How many 100s make 1,000? How many 10s make 1,000? How many 1s make 1,000? Can you see a pattern?*

ASSESSMENT CHECKPOINT Questions **1** and **2** will allow you to see whether children can count in 1,000s and what strategies they use. Can children use a number track effectively to find the missing numbers? Do children need equipment to help them with this?

ANSWERS

Question **1** a): There are 3,000 strawberry sweets.

Question **1** b): There are 5,000 lemon sweets.

Question **1** c): In total there are 8,000 sweets: 3,000 + 5,000 = 8,000.

Question **2** a): 0, 1,000, 2,000, 3,000, 4,000, 5,000, 6,000, 7,000

Question **2** b): 1,000, 2,000, 3,000, 4,000, 5,000

Question **2** c): 6,000, 7,000, 8,000, 9,000, 10,000

Question **2** d): 8,000, 7,000, 6,000, 5,000, 4,000, 3,000

Question **3** a): 10 hundreds = 1,000

Question **3** b): 100 tens = 1,000

Question **3** c): 1,000 ones = 1,000

Question **3** d): 20 hundreds = 2,000

Question **3** e): 200 tens = 2,000

Question **3** f): 2,000 ones = 2,000

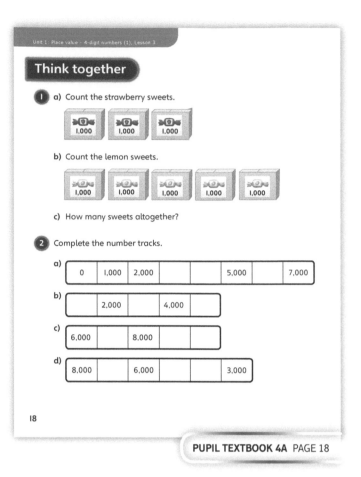

PUPIL TEXTBOOK 4A PAGE 18

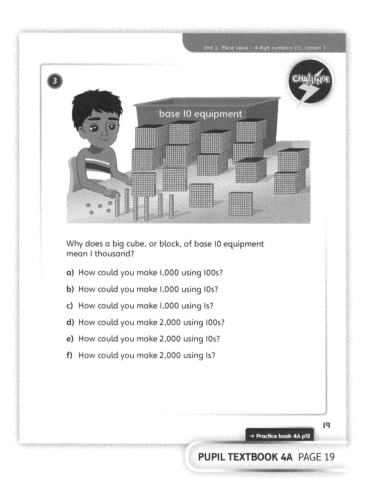

PUPIL TEXTBOOK 4A PAGE 19

Practice

WAYS OF WORKING Independent thinking

IN FOCUS
- Question ❶ further supports and develops children's ability to count in 1,000s.
- Question ❷ encourages children to use concrete equipment to recognise and count in 1,000s.
- Question ❸ allows children to use their previous learning of number tracks to count in 1,000s.

STRENGTHEN If children struggle to count in 1,000s, allow them to count in 100s to begin with, and then build on this understanding. Children can also use concrete resources to build 100s and count in 100s, then replicate this with 1,000s. Ensure that children are clearly linking the resources to counting, and count aloud with them.

DEEPEN Question ❻ challenges children to use their reasoning skills to identify multiples of 1,000. Encourage children to work in pairs to prove their answers.

ASSESSMENT CHECKPOINT Questions ❶, ❷ and ❸ will allow you to assess whether children can count in 1,000s, forwards and backwards, and whether they can continue a counting sequence regardless of the start point (i.e. not always starting from 0). Question ❺ should allow you to see if children can show what they have understood about counting in 1,000s. They should arrive at the answer by counting the number of red and blue pencils, and then counting the remaining boxes to see how many are green.

ANSWERS Answers for the **Practice** part of the lesson can be found in the *Power Maths* online subscription.

Reflect

WAYS OF WORKING Pair work

IN FOCUS This activity requires children to work with a partner to count up in multiples of 1,000. They take it in turns to say alternating multiples of 1,000 from 0 to 10,000.

ASSESSMENT CHECKPOINT Look for children counting forwards in 1,000s to assess their understanding of the lesson. Do they recognise the pattern of the count and correct any mistakes that their partner might make?

ANSWERS Answers for the **Reflect** part of the lesson can be found in the *Power Maths* online subscription.

After the lesson

- Were children confident counting in 1,000s?
- Are children able to count from any multiple of 1,000, forwards and backwards in 1,000s?
- Can children use concrete equipment, such as base 10, to count in 1,000s?

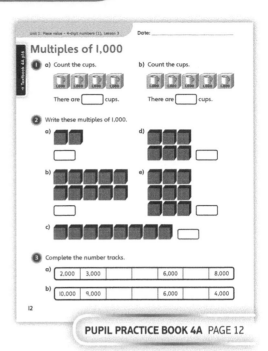

PUPIL PRACTICE BOOK 4A PAGE 12

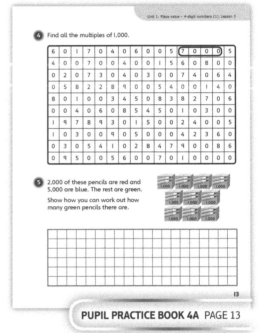

PUPIL PRACTICE BOOK 4A PAGE 13

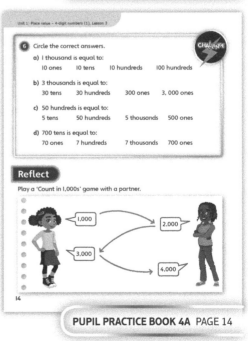

PUPIL PRACTICE BOOK 4A PAGE 14

4-digit numbers

Learning focus

In this lesson, children develop their understanding of place value by working with 4-digit numbers and understanding the place value of the 1,000s position.

Before you teach

- Can children count in 1,000s?
- Have they had the opportunity to rehearse the place value of 3-digit numbers?
- How could you help children to explore the place value of the columns in a place value grid showing 100s, 10s and 1s?

NATIONAL CURRICULUM LINKS

Year 4 Number – number and place value

Identify, represent and estimate numbers using different representations.

ASSESSING MASTERY

Children can recognise the value of each digit in a 4-digit number.

COMMON MISCONCEPTIONS

Children may make mistakes in using the language of 4-digit numbers, saying for example four thousand and two hundred and sixty-five. Ask:
- *How many 1,000s? What is the 100s, 10s and 1s part of the number? What do you say when you put all those bits together?*

STRENGTHENING UNDERSTANDING

Ask children to start with a 3-digit number, and then to place a digit before it to make a 4-digit number, as in the **Discover** question.

GOING DEEPER

Challenge children to find 4-digit numbers in science books or other sources of information such as animal fact-files.

KEY LANGUAGE

In lesson: thousands (1,000s), digit, 4-digit number

STRUCTURES AND REPRESENTATIONS

Place value grid

RESOURCES

Mandatory: place value equipment, dice

Optional: digit cards

 In the eTextbook of this lesson, you will find interactive links to a selection of teaching tools.

Quick recap

Ask children to take turns to roll three dice and to use the digits to make a 3-digit number. Repeat several times. Ask children to show the numbers on place value grids and then to describe them by saying:

My number has ? hundreds, ? tens and ? ones.

Discover

WAYS OF WORKING Pair work

ASK

- Question ① a): *Which three cards can you choose? What is the value of each digit in your 3-digit number?*
- Question ① b): *What will your new digit be worth? Can you make a prediction?*

IN FOCUS Children are building on their knowledge of 3-digit numbers to explore the place value of 4-digit numbers, in particular the thousands column.

PRACTICAL TIPS Use this as a chance for children to explore the language and make predictions, before giving an explanation of the value of the thousands position in a 4-digit number.

Use digit cards to do the activity together. Children make a 3-digit number, say the number, and then place a fourth digit in front. Rehearse how to say the 4-digit numbers including the correct language patterns.

ANSWERS

Question ① a): Various answers are possible, such as 124, 245, 521. Any combination of 3 digits. Children should show their number using base ten equipment in the correct order: 100s, 10s then 1s.

Question ① b): Various combinations are possible, with children adding the relevant number of 1,000 blocks to their base 10 equipment.

Share

WAYS OF WORKING Whole class teacher led

ASK

- Question ① a): *What is the value of each digit?*
- Question ① b): *How many thousands are in the thousands place?*

IN FOCUS Use question ① a) to help children rehearse the concept of place value in 3-digit numbers, and then move on to question ① b) where they begin to understand the value of the thousands place in 4-digit numbers.

PUPIL TEXTBOOK 4A PAGE 20

PUPIL TEXTBOOK 4A PAGE 21

Think together

WAYS OF WORKING Whole class teacher led (I do, We do, You do)

ASK

- Question **1**: *What is the value of each digit?*
- Question **2**: *What is the same and what is different about each of these 4-digit numbers?*
- Question **3**: *How might a '0' make things trickier for you?*

IN FOCUS In question **1** children recognise and write 4-digit numbers that are shown with place value counters. In question **2** they will need to understand how to say numbers that include zero as a place holder. Question **3** explores variations in the digits of 4-digit numbers.

STRENGTHEN Provide digit cards and place value grids to support children's use of the language of place value.

DEEPEN Ask children to investigate 4-digit numbers that they can find in real-life or practical contexts. What can they say about the numbers that they find?

ASSESSMENT CHECKPOINT Use question **2** to assess whether children recognise how to say each number and can explain the place value of each digit, including zero digits.

ANSWERS

Question **1**: Three thousand, four hundred and sixty-two; 3,462

Four thousand, two hundred and fifty-six; 4,256

Question **2**: Three thousand, two hundred and fifty; 3 thousands, 2 hundreds, 5 tens

Three thousand, two hundred and five; 3 thousands, 2 hundreds, 5 ones

Three thousand and twenty-five; 3 thousands, 2 tens, 5 ones

Question **3**: All numbers must be 4 digits, so 0 cannot be in the thousands place. There are 9 possible solutions: 1022, 1202, 1220, 2012, 2021, 2102, 2120, 2201, 2210.

PUPIL TEXTBOOK 4A PAGE 22

PUPIL TEXTBOOK 4A PAGE 23

Practice

WAYS OF WORKING Independent thinking

IN FOCUS In question ① children recognise different place value representations and match pictorial to abstract representations. Question ② uses place value grids to show 4-digit numbers, whilst question ③ asks children to draw their own representations of place value equipment. Question ④ uses digit cards to explore possible variations in the digits of 4-digit numbers and question ⑤ requires children to use their knowledge of place value in 4-digit numbers to solve a logic puzzle.

STRENGTHEN Rehearse drawing place value representations by giving children more questions in the style of question ③. Provide place value equipment for children to use alongside.

DEEPEN Challenge children to create their own 4-digit place value logic puzzles with a mystery number like the one in question ⑤.

ASSESSMENT CHECKPOINT Use question ② to assess whether children can correctly interpret place value representations and use them to write 4-digit numbers, showing an understanding of the value of each digit.

ANSWERS Answers for the **Practice** part of the lesson can be found in the *Power Maths* online subscription.

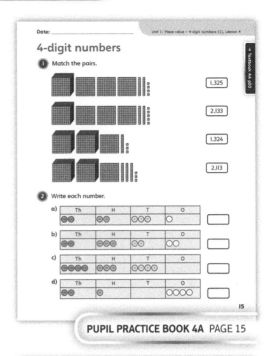

PUPIL PRACTICE BOOK 4A PAGE 15

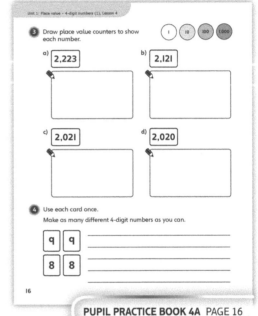

PUPIL PRACTICE BOOK 4A PAGE 16

Reflect

WAYS OF WORKING Independent thinking

IN FOCUS The **Reflect** part of the lesson allows children to explain the place value of each digit in a 4-digit number, and to explore the possible variations that could occur in one of the digits.

ASSESSMENT CHECKPOINT Assess whether children can explain the place value of each digit in a 4-digit number and use this to justify their answer.

ANSWERS Answers for the **Reflect** part of the lesson can be found in the *Power Maths* online subscription.

PUPIL PRACTICE BOOK 4A PAGE 17

After the lesson ⏸

- Write some 4-digit numbers on the board and ask children to discuss the place value of each digit.
- Can children draw or make each number with place value equipment?

Partition 4-digit numbers

Learning focus

In this lesson, children further explore the value of each digit in a 4-digit number by partitioning into 1,000s, 100s, 10s and 1s.

Before you teach

- Have children had plenty of practise writing and reading 4-digit numbers?
- Can children represent 4-digit numbers using a variety of equipment?
- Do children know how to partition 2- and 3-digit numbers?

NATIONAL CURRICULUM LINKS

Year 4 Number – number and place value

Recognise the place value of each digit in a four-digit number (1,000s, 100s, 10s, and 1s).

ASSESSING MASTERY

Children can partition and recombine 4-digit numbers into 1,000s, 100s, 10s and 1s.

COMMON MISCONCEPTIONS

Children may not know how to partition numbers where 0 is used as a place holder. Ask:
- *What is the value of each digit?*
- *How many thousands are there in this number? How many 100s? 10s? 1s?*

STRENGTHENING UNDERSTANDING

Provide place value equipment to support each place value addition, so that children can physically see how to separate and recombine the partitions for a given number.

GOING DEEPER

Ask children to explore the concept of 1 kg as 1,000 g, in order to understand the role of place value in metric units.

KEY LANGUAGE

In lesson: partition, part, whole, thousands (1,000s), hundreds (100s), tens (10s), ones (1s)

STRUCTURES AND REPRESENTATIONS

Part-whole model, place value grid

RESOURCES

Mandatory: place value equipment, counters

 In the eTextbook of this lesson, you will find interactive links to a selection of teaching tools.

Quick recap

Write these 3-digit numbers on the board:

123, 320, 301, 231, 222

Ask children to use place value equipment to make each number.

Discover

Pair work

ASK

- Question **1** a): *What is the same and what is different about how you count the legs on each of the creatures?*
- Question **1** b): *How do you know you've counted the correct number of 1,000s / 100s / 10s / 1s?*

IN FOCUS Children are presented with the context of creatures with many legs as a prompt for counting in steps of different unitised amounts. The creatures represent 1,000s, 100s, 10s and 1s respectively.

PRACTICAL TIPS Count together as a class, rehearsing the patterns that can be seen when counting in 1s, then in 10s, then in 100s and then in 1,000s.

ANSWERS

Question **1** a): 4 Thods = 4,000 legs; 5 Hods = 500 legs;
3 Tods = 30 legs; 1 Od = 1 leg

Question **1** b): 4,000 + 500 + 30 + 1 = 4,531 legs altogether.

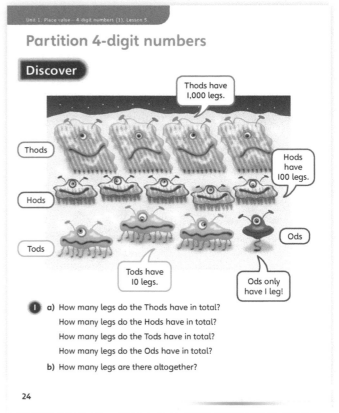

PUPIL TEXTBOOK 4A PAGE 24

Share

Whole class teacher led

ASK

- Question **1** a): *How can we check? Have you counted the number of creatures in each set?*
- Question **1** b): *Why does it not say 4 + 5 + 3 + 1 = 4,531?*

IN FOCUS Question **1** a) establishes children's understanding of the units of place value for 4-digit numbers: 1,000s, 100s, 10s and 1s. In question **1** b) children are then using the given context to partition the total number of legs into 1,000s, 100s, 10s and 1s. This is shown using a variety of representations.

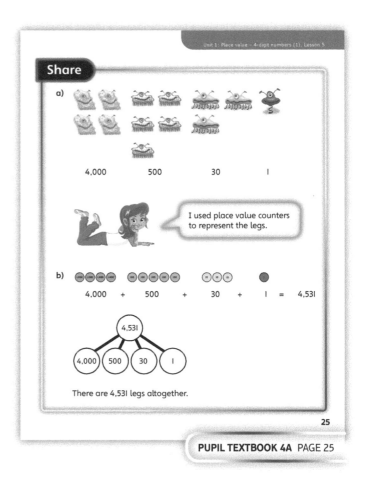

PUPIL TEXTBOOK 4A PAGE 25

Think together

WAYS OF WORKING Whole class teacher led (I do, We do, You do)

ASK

- Question **1**: *What is the value of each digit? What is the whole? What are the parts?*
- Question **2**: *Do the parts always appear in the same order?*
- Question **3**: *Why does a '0' sometimes make things trickier? What advice would you give?*

IN FOCUS Question **1** provides scaffolding to help children understand the link between partitioning and place value additions. Children use this in question **2** to partition and combine using addition. Question **3** requires children to understand the role of '0' as a place holder.

STRENGTHEN Ask children to check their answers to partitioning questions by using place value equipment alongside their work.

DEEPEN Challenge children to find several different ways to write the partitions of a given number, by writing the parts in a different order each time. For example:

3,254 = 3,000 + 200 + 50 + 4

3,254 = 3,000 + 4 + 50 + 200

ASSESSMENT CHECKPOINT Use question **2** to assess whether children can identify the missing numbers in place value additions, including where the partitions are presented in a different order.

ANSWERS

Question **1**: 2 thousands, 5 hundreds, 4 tens, 2 ones = 2,000 + 500 + 40 + 2

Question **2** a): 100, 5

Question **2** b): 3,275

Question **2** c): 2,651

Question **2** d): 200

Question **3** a): **5,**000   **200**   **10**
                5,000   **200**   **1**
                5,000   **70**

Question **3** b): **3,033** = 3,000 + 30 + 3
               3,303 = 3 + 300 + 3,000
               9,009 = 9,000 + 9
               9,090 = 90 + 9,000
               7,070 = **7,000** + **0** + **70** + **0**
               7,707 = **7,000** + **700** + **0** + **7**

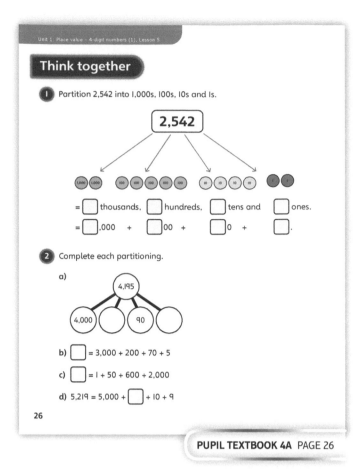

PUPIL TEXTBOOK 4A PAGE 26

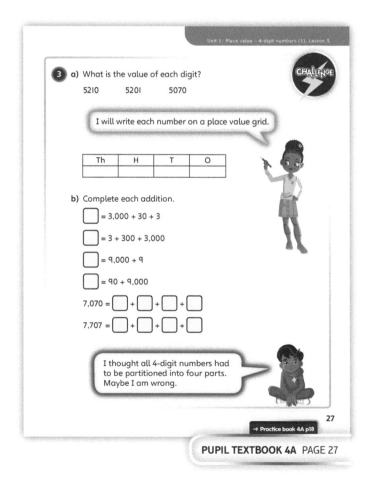

PUPIL TEXTBOOK 4A PAGE 27

Practice

WAYS OF WORKING Independent thinking

IN FOCUS In question ① children must recognise the place value of each digit in 4-digit numbers and in question ② they complete place value additions to show partitions. Question ③ requires children to recognise the value of certain digits. When working through questions ④ and ⑤, children need to understand the role of '0' as a place holder.

STRENGTHEN Provide place value equipment for children to use alongside their partitioning work. Encourage them to show you a representation of each number and to explain what this tells them before finding an answer.

DEEPEN In question ②, children explore possible variations in the digits of 4-digit numbers. Show them two possible partitions for the number 2,357, for example:

2,357 = 2,000 + 300 + 50 + 7
2,357 = 7 + 50 + 300 + 2,000

Explain that there are 22 more ways to write the partitions of 2,357 in 1,000s, 100s, 10s and 1s. Can children find them all?

ASSESSMENT CHECKPOINT Assess whether children can complete question ④ accurately, demonstrating that they recognise the value of each digit and understand the role of '0' as a place holder.

ANSWERS Answers for the **Practice** part of the lesson can be found in the *Power Maths* online subscription.

Reflect

WAYS OF WORKING Pair work

IN FOCUS The **Reflect** part of the lesson asks children to write a number puzzle in the style of the Challenge question in the **Practice** section. They should use their knowledge of place value and partitioning of 4-digit numbers to write appropriate clues.

ASSESSMENT CHECKPOINT Assess whether children can correctly use their knowledge of the place value of 4-digit numbers to accurately write clues and justify their working.

ANSWERS Answers for the **Reflect** part of the lesson can be found in the *Power Maths* online subscription.

After the lesson

- Can children partition and recombine a set of 4-digit numbers in a variety of ways?
- Could children use their place value knowledge to write more puzzles for each other?
- How else could you explore the role of '0' together?

Unit 1: Place value – 4-digit numbers (1), Lesson 5   Date: _____

Partition 4-digit numbers

① a) Partition each number into thousands, hundreds, tens and ones.

2,324 = ☐ thousands, ☐ hundreds, ☐ tens and ☐ ones

6,281 = ☐ thousands, ☐ hundreds, ☐ tens and ☐ ones

4,427 = ☐ thousands, ☐ hundreds, ☐ tens and ☐ ones

9,988 = ☐ thousands, ☐ hundreds, ☐ tens and ☐ ones

b) Complete each number

☐ = 5 thousands, 2 hundreds, 3 tens and 7 ones

☐ = 2 thousands, 8 hundreds, 9 tens and 4 ones

☐ = 9 thousands, 1 hundred, 3 tens and 6 ones

☐ = 7 thousands, 6 hundreds, 5 tens and 4 ones

② Complete each partition as an addition.

a) ☐ = 3,000 + 500 + 10 + 1

b) ☐ = 5,000 + 300 + 90 + 3

c) ☐ = 5 + 30 + 900 + 7,000

d) ☐ = 9,000 + 7 + 50 + 300

e) 1,574 = 4 + 70 + ☐ + 1,000

f) 4,141 = 1 + 40 + 100 + ☐

18

PUPIL PRACTICE BOOK 4A PAGE 18

Unit 1: Place value – 4-digit numbers (1), Lesson 5

③ Use a tick to show the value of each underlined digit.

	5	50	500	5,000
2,5̲52				
5,235̲				
1,5̲55				
5̲,055				

④ Join matching pairs.

2,068		2,000 + 800 + 6
2,608		6,000 + 800 + 2
2,806		2,000 + 60 + 8
2,680		2,000 + 600 + 80
6,820		6,000 + 800 + 20
6,802		2,000 + 600 + 8

⑤ Partition each number into place value additions.

a) 4,400 = _____   d) 3,030 = _____

b) 4,040 = _____   e) 1,010 = _____

c) 4,004 = _____   f) 6,060 = _____

19

PUPIL PRACTICE BOOK 4A PAGE 19

Unit 1: Place value – 4-digit numbers (1), Lesson 5

⑥ Andy has made a number. He says:

CHALLENGE

- My number has the same number of 1,000s and 10s.
- There are two more 1s than 10s.
- The 100s digit is half the 1,000s digit.

What could Andy's number be?

Draw place value counters to show the possible answers.

Reflect

Make up your own mystery number puzzle.

Challenge a partner to solve it.

○ _____
○ _____
○ _____
○ _____
○ _____

20

PUPIL PRACTICE BOOK 4A PAGE 20

Partition 4-digit numbers flexibly

Learning focus

In this lesson, children explore partitioning 4-digit numbers in various ways, not necessarily just into 1,000s, 100s, 10s and 1s. This prepares them for conceptual work with written calculation methods.

Before you teach

- Do children know how to make and represent 4-digit numbers using place value equipment?
- Can children partition 4-digit numbers into 1,000s, 100s, 10s and 1s and also recombine 1,000s, 100s, 10s and 1s into 4-digit numbers?

NATIONAL CURRICULUM LINKS

Year 4 Number – number and place value

Recognise the place value of each digit in a four-digit number (1,000s, 100s, 10s, and 1s).

Identify, represent and estimate numbers using different representations.

ASSESSING MASTERY

Children can complete partitions of 4-digit numbers in a variety of ways, beyond just 1,000s, 100s, 10s and 1s. For example, $4,235 = 400 + 200 + 20 + \boxed{}$.

COMMON MISCONCEPTIONS

Children may think that partitioning of a 4-digit number can only be done into the units of 1,000s, 100s, 10s and 1s. Ask:
- *Can you make a 4-digit number from place value equipment and then partition it? Can you split it into different parts? Is there another way? And another way?*

STRENGTHENING UNDERSTANDING

Provide place value equipment for children to use alongside 4-digit numbers, so that children can physically move the equipment into different groups.

GOING DEEPER

Ask children to investigate the link between partitioning and addition by exploring questions such as: *How could thinking about flexible partitioning help us to solve additions such as 2,000 + 1,200?*

KEY LANGUAGE

In lesson: part, whole, partition

Other language to be used by the teacher: addition, subtraction

STRUCTURES AND REPRESENTATIONS

Part-whole model

RESOURCES

Mandatory: place value equipment

 In the eTextbook of this lesson, you will find interactive links to a selection of teaching tools.

Quick recap

Write these 4-digit numbers on the board:

1,123, 2,320, 1,301, 2,231, 2,222

Ask children to use place value equipment to make each number.

Discover

WAYS OF WORKING Pair work

ASK

- Question ❶: *What do you notice about all the masses?*
- Question ❶ a): *How could you organise the parcels?*
- Question ❶ b): *How many parcels are there? How many sacks are there?*

IN FOCUS This gives children the visual cue of five masses representing various place values being sorted into only three groups, to introduce the idea of partitioning 4-digit numbers more flexibly than just into 1,000s, 100s, 10s and 1s.

PRACTICAL TIPS Ask children to use place value equipment to represent the mass of each parcel. Can they describe how the equipment can be rearranged in a variety of ways?

ANSWERS

Question ❶ a): There are several ways to divide the parcels and letter into three sacks:

$2 \times 1{,}000$ g + 200 g + **50 g + 5 g**

$1{,}000$ g + **1,000 g + 200 g** + 50 g + 5 g

$1{,}000$ g + 1,000 g + **200 g + 50 g + 5 g**

Question ❶ b): The total mass is 2,255 g.

Unit 1: Place value – 4-digit numbers (1), Lesson 6

Partition 4-digit numbers flexibly

Discover

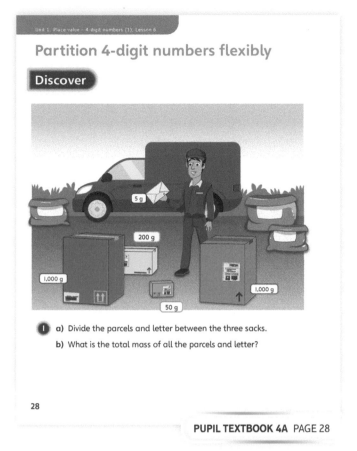

❶ a) Divide the parcels and letter between the three sacks.

b) What is the total mass of all the parcels and letter?

28

PUPIL TEXTBOOK 4A PAGE 28

Share

WAYS OF WORKING Whole class teacher led

ASK

- Question ❶ a): *Can you think of any other ways to sort the parcels?*
- Question ❶ b): *What is the same and what is different about each part-whole model?*

IN FOCUS Question ❶ a) demonstrates various ways to arrange and sort the different parts. How many other ways can children find? In question ❶ b) children then recombine the parts into a whole. They should find that however they have sorted the parts, the whole is always the same.

Share

a) Here are two ways you can divide the parcels and letter.

b) The total mass is the same, however you sort them.

There are lots of ways I could have put the parcels and letter into the sacks.

29

PUPIL TEXTBOOK 4A PAGE 29

Think together

ASK

- Question **1**: *Could you partition this number into three parts? What about two parts?*
- Question **2**: *What are the parts here? What are the wholes?*
- Question **3**: *Which part is being subtracted each time?*

IN FOCUS Question **1** shows children an example of flexible partitioning, which they can then continue to explore. In question **2**, children complete partitioning additions, including finding missing parts. When children reach question **3**, they can use their knowledge of flexible partitioning to solve calculations.

STRENGTHEN Ensure children use place value equipment alongside each question to support their reasoning and check their answers.

DEEPEN Develop the use of partitioning in different ways like this to support children's calculation thinking for addition and subtraction. Give a range of additions and subtractions and ask children to discuss how flexible partitioning could help them to solve each one. Are there any calculations where using flexible partitioning is not the most useful method?

ASSESSMENT CHECKPOINT Use question **2** to assess whether children can successfully complete flexible partitioning calculations with missing numbers, including finding parts and wholes.

ANSWERS

Question **1**: There are multiple ways to partition 3,225, including partitioning each digit:
$3{,}000 + 225$
$2{,}100 + 1{,}125$
$3{,}200 + 20 + 5$
$3{,}000 + 210 + 15$

Question **2** a): 1,435

Question **2** b): 2,575

Question **2** c): 2,236

Question **2** d): 85

Question **2** e): 14

Question **2** f): 2,500

Question **3** a): 5,075   5
          5,705   5,000

Question **3** b): 5,760
          5,715
          1,710
          4,060

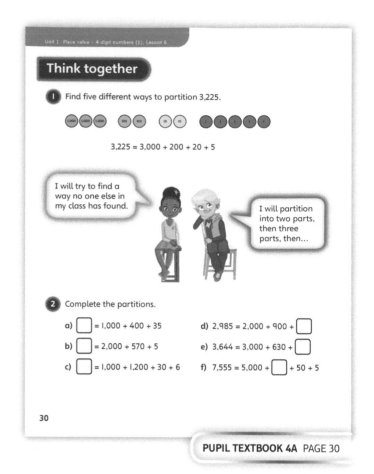

PUPIL TEXTBOOK 4A PAGE 30

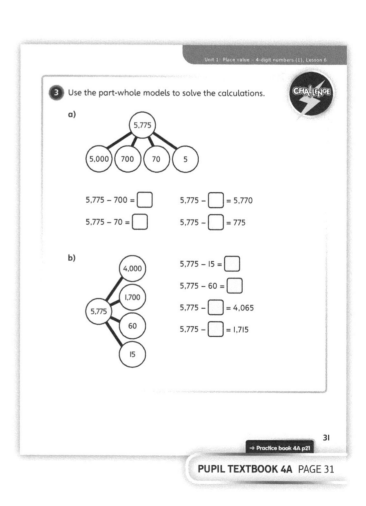

PUPIL TEXTBOOK 4A PAGE 31

Practice

WAYS OF WORKING Independent thinking

IN FOCUS In question ❶, children partition a 4-digit number flexibly in several different ways. In question ❷ they complete partition additions with missing numbers, where they must find both parts and wholes. Questions ❸, ❺ and ❻ require children to use partitioning techniques to solve calculations. Question ❹ focuses specifically on subtraction.

STRENGTHEN Ask children to make predictions about the partitions and to then use place value equipment to check their thinking.

DEEPEN How many different partitions can children find for the number 2,405?

ASSESSMENT CHECKPOINT Question ❷ provides the core learning for this lesson. Assess whether children can explain how they will use the information given and their understanding of flexible partitioning to find each of the missing numbers.

ANSWERS Answers for the **Practice** part of the lesson can be found in the *Power Maths* online subscription.

Reflect

WAYS OF WORKING Pair work

IN FOCUS The **Reflect** part of the lesson requires children to use flexible partitioning with a number that has '0' as a place holder digit.

ASSESSMENT CHECKPOINT Assess whether children can find more than one partition and can justify their answers.

ANSWERS Answers for the **Reflect** part of the lesson can be found in the *Power Maths* online subscription.

After the lesson ⏸

- Can children find five different partitions for any given 4-digit number?

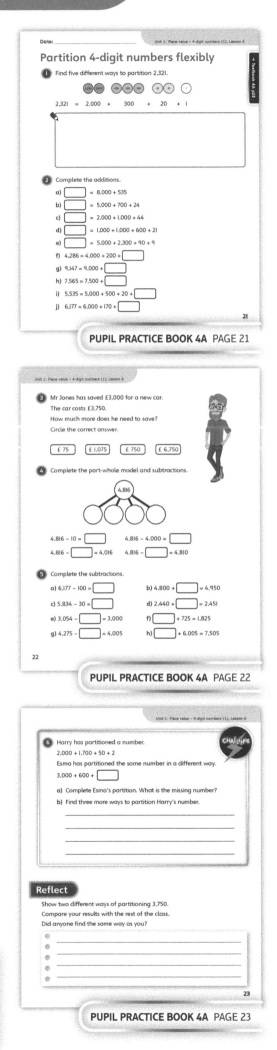

PUPIL PRACTICE BOOK 4A PAGE 21

PUPIL PRACTICE BOOK 4A PAGE 22

PUPIL PRACTICE BOOK 4A PAGE 23

1, 10, 100, 1,000 more or less

Learning focus

In this lesson, children will find 1,000 more or less than a given number, using their knowledge of place value to help them. Children will also recap their learning on 10 and 100 more.

Before you teach

- Can children find 1 more or less than a number?
- Can children represent numbers using concrete equipment?

NATIONAL CURRICULUM LINKS

Year 3 Number – number and place value

Count from 0 in multiples of 4, 8, 50 and 100; find 10 or 100 more or less than a given number.

Year 4 Number – number and place value

Find 1,000 more or less than a given number.

ASSESSING MASTERY

Children can successfully find 1, 10, 100 and 1,000 more or less than a given number in a range of contexts and identify which place value column will help them to do this. Children can use a variety of concrete apparatus to demonstrate their understanding.

COMMON MISCONCEPTIONS

When presented with an abstract number, children may not be able to distinguish between finding 1, 10, 100 or 1,000 more or less than the number. To reinforce children's place value understanding, and to help you address any misconceptions, ask:
- *Can you check your answer using base 10 equipment and a place value grid?*

STRENGTHENING UNDERSTANDING

Provide children with opportunities to find 1, 10 and 100 more or less than different numbers, using concrete equipment, such as base 10 equipment, to support them. Next, progress to 1,000 more or less than a number. Ensure children's understanding of how to use base 10 equipment, and what each cube represents, is secure. For example, give children a number such as 3,264 and ask them to say out loud how many 1,000s, 100s, 10s and 1s make up this number. Use a place value grid to further support understanding, asking children to complete a grid for each 4-digit number you discuss. Can they find 1, 10, 100 or 1,000 more or less than each number, using the place value grid to support them?

GOING DEEPER

Can children find 10 or 100 more than a given number that might involve using an exchange? For example, can children work out 10 less than 2,003 or 100 less than 6,004? Encourage children to explore and think deeply about which digits change if they find 10, 100 or 1,000 more or less than a given number. Ask: *When you are finding 10 more or less, can the 1,000s digit change? Give an example of when this might happen.*

KEY LANGUAGE

In lesson: less than, more than, subtract, exchange, place value, tens (10s), hundreds (100s), thousands (1,000s), function machine, step

Other language to be used by the teacher: more, less, greater, smaller, greatest, smallest, least, base 10 equipment, place value counters, place value grids

STRUCTURES AND REPRESENTATIONS

Place value grid

RESOURCES

Mandatory: place value counters, base 10 equipment

Optional: number lines, place value grids, base 10 equipment

 In the eTextbook of this lesson, you will find interactive links to a selection of teaching tools.

Quick recap

Count on together in steps of 1 from a given 2-digit number. Then, count on in steps of 10 from a given 3-digit number. Finally, count on in steps of 100 from a given 4-digit number.

Now reverse it and count down together in steps of 1 from a given 2-digit number. Then count down in steps of 10 from a given 3-digit number. Finally, count down in steps of 100 from a given 4-digit number.

Discover

Unit 1: Place value – 4-digit numbers (1), Lesson 7

WAYS OF WORKING Pair work

ASK

- Question **1** a): *What does the word 'attendance' mean? How many people attended the game?*
- Question **1** a): *What does each digit in the attendance number represent?*
- Question **1** b): *Which place value column changes when you have 1,000 more? Do any other columns change? Why?*

IN FOCUS Question **1** a) builds on children's understanding of place value. Encourage children to support their thinking by saying aloud how many 1,000s, 100s, 10s and 1s are in the number as they are making it.

Question **1** b) extends work children have done in previous years, but now dealing with 4-digit numbers. Children should focus on the 1,000s column and look at how this changes when 1,000 is added.

PRACTICAL TIPS Allow children to use concrete equipment or a number line to help them visualise the numbers they are working with.

ANSWERS

Question **1** a): Children should have 8 thousands, 9 hundreds, 8 tens and 0 ones.

Question **1** b): 1,000 more than 8,980 is 9,980. The thousands columns should now have 9 counters.

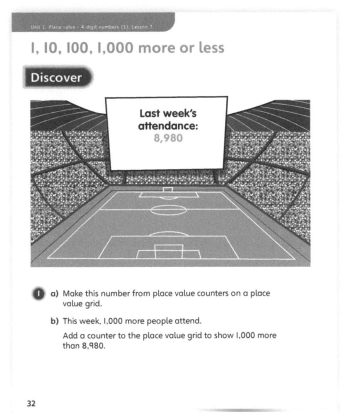

1, 10, 100, 1,000 more or less

Discover

Last week's attendance: 8,980

1 a) Make this number from place value counters on a place value grid.

b) This week, 1,000 more people attend.
Add a counter to the place value grid to show 1,000 more than 8,980.

32

PUPIL TEXTBOOK 4A PAGE 32

Share

WAYS OF WORKING Whole class teacher led

ASK

- Question **1** a): *What does each digit represent? How does this link to the base 10 equipment? How does a place value grid help here?*
- Question **1** b): *When finding 1,000 more, which place value counter do you need to add? Which column does it go in?*

IN FOCUS In question **1** a), children must represent a given 4-digit number with concrete place value equipment. Using base 10 equipment could also support this.

In question **1** b) the focus is on adding 1,000. Encourage children to identify which counter they need to add and where it will go, and to describe how many counters are now in each column in order to find the final number.

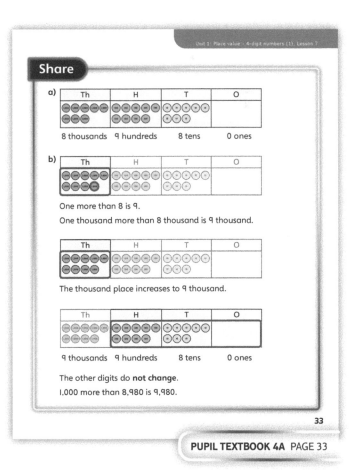

Share

a)

Th	H	T	O
8 thousands	9 hundreds	8 tens	0 ones

b)

Th	H	T	O

One more than 8 is 9.
One thousand more than 8 thousand is 9 thousand.

Th	H	T	O

The thousand place increases to 9 thousand.

Th	H	T	O
9 thousands	9 hundreds	8 tens	0 ones

The other digits do **not change**.
1,000 more than 8,980 is 9,980.

33

PUPIL TEXTBOOK 4A PAGE 33

Think together

Unit 1: Place value – 4-digit numbers (1), Lesson 7

Think together

WAYS OF WORKING Whole class teacher led (I do, We do, You do)

ASK

- Question **1**: *What would you change to show 10 more? 10 less?*
- Question **2**: *When representing 100 less than 7,892 on a place value grid, which place value column will change?*
- Question **3** c): *Which number do you need to work out here? What are some ways you could work out this number?*

IN FOCUS Question **2** shows a 4-digit number represented with place value counters. Children explore how each place value column changes when finding 1,000, 100, 10 and 1 less.

Question **3** consolidates learning on finding 10, 100 and 1,000 more or less than a given number using a fun function machine. Question **3** c) requires children to think a little differently.

STRENGTHEN To support children in adding and subtracting 1, 10, 100 and 1,000 to or from a 4-digit number, encourage them to make the number with base 10 equipment on a place value grid, and then add or remove the relevant pieces of equipment to ensure they clearly understand what is happening.

DEEPEN Ask children to think deeply about place value with questions such as: *Which columns can change when finding 10, 100 and 1,000 more than a given number?*

Question **3** c) explores that 10 more is the opposite of 10 less, reinforced by 100 and 1,000 more or less.

ASSESSMENT CHECKPOINT Assess children's mastery of 1, 10, 100 and 1,000 more or less by asking them to find 1, 10, 100 or 1,000 more or less than a variety of 4-digit numbers. Look for fluency and consistency; are there children who would benefit from further support to help embed understanding?

ANSWERS

Question **1** a): Children should now have 6 hundreds counters.

Question **1** b): The grid should be back to the starting number.

Question **1** c): Adding 100 then subtracting 100 returns it to the starting number.

Question **2** a): 6,892

Question **2** b): 7,792

Question **2** c): 7,882

Question **2** d): 7,891

Question **2** e): 8,892

Question **2** f): 7,992

Question **2** g): 7,902

Question **2** h): 7,893

Question **3** a): 6,785

Question **3** b): 8,103

Question **3** c): 6,540

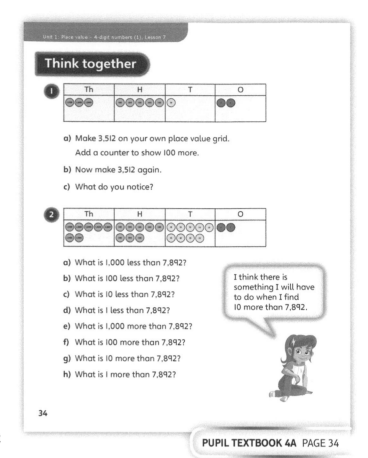

1

a) Make 3,512 on your own place value grid.
   Add a counter to show 100 more.

b) Now make 3,512 again.

c) What do you notice?

2

a) What is 1,000 less than 7,892?

b) What is 100 less than 7,892?

c) What is 10 less than 7,892?

d) What is 1 less than 7,892?

e) What is 1,000 more than 7,892?

f) What is 100 more than 7,892?

g) What is 10 more than 7,892?

h) What is 1 more than 7,892?

I think there is something I will have to do when I find 10 more than 7,892.

34

PUPIL TEXTBOOK 4A PAGE 34

3 Here is a function machine.

IN

1,000 more | 100 more | 10 more

OUT

a) What number comes out when 5,675 is put in?

b) What number comes out when 6,993 is put in?

c) If 7,650 comes out, what number was put in?

I will work out the answer at each step so I do not make mistakes.

I wonder if I would get the same answer if the functions were in a different order.

35

→ Practice book 4A p24

PUPIL TEXTBOOK 4A PAGE 35

Practice

Independent thinking

IN FOCUS Question ❶ gives children further practice finding 10, 100 and 1,000 more or less than a number, using pictorial models to support them. Children may want to use concrete equipment to further support their work in this question.

Question ❷ focuses on children's understanding of a variety of pictorial representations. Use this question to look for any misconceptions in children's understanding of these models.

Questions ❸ and ❹ ask children to find the missing numbers to complete some number sentences. The questions slowly build in difficulty and key digits have been kept the same to encourage children to spot patterns as they work through their answers.

STRENGTHEN Reinforce place value understanding by reading the numbers aloud and asking children to build each digit of each number using base 10 equipment. Then ask them to add or take away the relevant amount. Ask: *What is the new number? Which digits have changed?*

DEEPEN Ask children to think creatively and explore examples of numbers in which, when they find 100 more than the number, a column other than the hundreds column also changes. For example, *What is 100 more than 4,982?* Similarly, can they find examples of numbers in which, when they find 1,000 more than the number, a column other than the thousands column also changes?

ASSESSMENT CHECKPOINT Question ❺ provides children with an opportunity to show that they can find 10, 100 and 1,000 more or less than a number, and apply the logic that they must work backwards from a given output number to find the original input number. You could use this question as a model to work through with the class, looking for and discussing any misconceptions as you do so.

ANSWERS Answers for the **Practice** part of the lesson can be found in the *Power Maths* online subscription.

Reflect

WAYS OF WORKING Pair work

IN FOCUS Children should recognise that it is the 1,000s digit that changes when finding 1,000 more or less and that the other digits stay the same. Explore finding 100 more or less with children, using examples to establish the fact that the 1,000s digit may also change.

ASSESSMENT CHECKPOINT This question assesses whether children can find 10, 100 and 1,000 more or less than a number. Do children know which numbers to focus on and which digits can change?

ANSWERS Answers for the **Reflect** part of the lesson can be found in the *Power Maths* online subscription.

After the lesson ⏸

- Can children use concrete equipment to explain what happens when finding 1,000 more or less than a given number?
- Can children find 1, 10 and 100 more or less than a given 4-digit number?
- Can children explain which place value columns change when finding 1, 10, 100 and 1,000 more or less, and why?

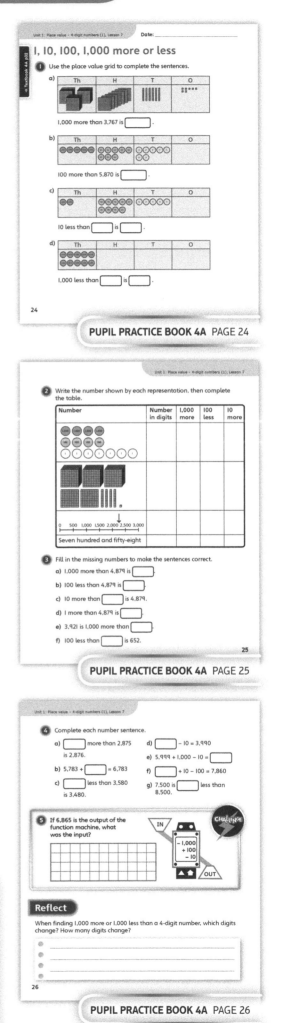

PUPIL PRACTICE BOOK 4A PAGE 24

PUPIL PRACTICE BOOK 4A PAGE 25

PUPIL PRACTICE BOOK 4A PAGE 26

1,000s, 100s, 10s and 1s

Learning focus

In this lesson, children develop their understanding of the relationship between 1,000s, 100s, 10s and 1s, and explore the concept of exchange more fully.

Before you teach 🔢

- Can children count in 10s and in 100s?
- Do children know how to use place value equipment to make numbers from 10s or 100s?

NATIONAL CURRICULUM LINKS

Year 4 Number – number and place value

Recognise the place value of each digit in a four-digit number (1,000s, 100s, 10s, and 1s).

Identify, represent and estimate numbers using different representations.

ASSESSING MASTERY

Children can convert numbers such as 1,400 into 14 hundreds or 140 tens.

COMMON MISCONCEPTIONS

Children may think that the idea of having '13 hundreds' is now 'allowed', without realising that you can still only have up to 9 in each place value column. Ask:

- Count in 10s up to 90. Can you continue the count? What happens after 100?

STRENGTHENING UNDERSTANDING

Use place value counters on ten frames to support understanding, linking counting in 10s and 100s to the numbers created. Explore the concept of 10 hundreds making 1,000.

GOING DEEPER

Challenge children to explore the different ways that they can make 9,900 out of 100s or 10s.

KEY LANGUAGE

In lesson: thousands (1,000s), hundreds (100s), tens (10s), ones (1s); number names such as 'fourteen hundred'

Other language to be used by the teacher: place value, digit

STRUCTURES AND REPRESENTATIONS

Ten frames

RESOURCES

Mandatory: counters, base 10 equipment, place value counters

 In the eTextbook of this lesson, you will find interactive links to a selection of teaching tools.

Quick recap 🔁

As a class, count in 10s from 0 to 200. Then count up in 100s from 0 to 2,000. Now reverse the counts to count down in steps of 10 and 100.

Discover

WAYS OF WORKING Pair work

WAYS OF WORKING Pair work

ASK

- Question ❶: *Have you ever heard numbers like 'fifteen hundred metres'?*
- Question ❶: *Why might some people think it sounds like an unusual number?*

IN FOCUS Children are exploring 100s in numbers, in preparation for exchanging between 10 hundreds and 1 thousand.

PRACTICAL TIPS Provide place value equipment to support children's thinking and to help them visualise the relationship between 1,000s and 100s.

ANSWERS

Question ❶ a): Children should show 15 hundreds squares.

Question ❶ b): Children should show 1 thousands block and 5 hundreds squares.

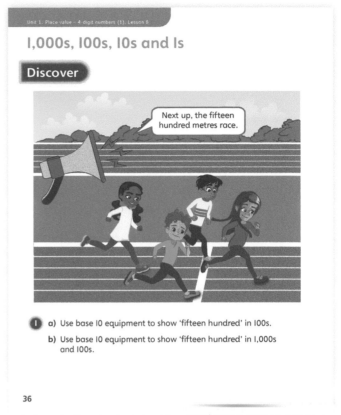

PUPIL TEXTBOOK 4A PAGE 36

Share

WAYS OF WORKING Whole class teacher led

ASK

- Question ❶ a): *Count in 100s. What number normally comes after 900 in the count? Can you think of another way to say this number?*
- Question ❶ b): *Why have 10 hundreds been swapped for 1,000? Is this mathematically correct?*

IN FOCUS Question ❶ a) supports children in learning to count beyond 9 hundreds to reach 10 hundred, 11 hundred, 12 hundred and so on. In question ❶ b) children make the exchange of 10 hundreds for 1 thousand and explore how this works in a number that is made up of more than 10 hundreds.

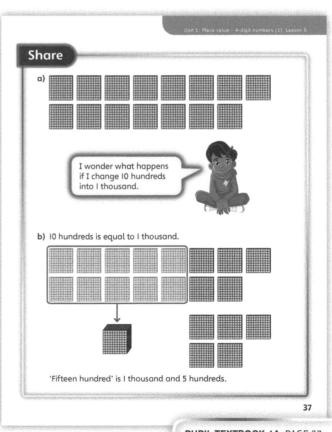

PUPIL TEXTBOOK 4A PAGE 37

Think together

WAYS OF WORKING Whole class teacher led (I do, We do, You do)

ASK

- Question **1**: *What is the same and what is different about each number? How do the ten frames help you see how to say the number?*
- Question **2**: *How will you arrange the place value counters?*
- Question **3**: *Can you predict the correct number without using all of the equipment?*

IN FOCUS In question **1** children are using ten frames to help them visualise how to convert 100s into 4-digit numbers. For question **2**, children make the given numbers themselves using place value counters on ten frames, and then write the numbers as digits. Question **3** requires children to write a given 4-digit number (a multiple of 100) as a number of 100s, and also as a number of 10s.

STRENGTHEN Use place value equipment and ten frames to make the numbers.

DEEPEN Ask children to convert the following numbers into 10s: 2,300 = ☐ tens; 3,300 = ☐ tens; 5,300 = ☐ tens. What do they notice?

ASSESSMENT CHECKPOINT Children can explain how numbers such as 25 hundred and 2,500 are equal to each other.

ANSWERS

Question **1**: 13 hundreds are equal to **1,300**.
29 hundreds are equal to **2,900**.

Question **2**: 25 hundreds = 2,500; 2 thousands, 5 hundreds
14 hundreds = 1,400; 1 thousand, 4 hundreds

Question **3**: 25 × 100 g = 2,500 g
250 × 10 g = 2,500 g

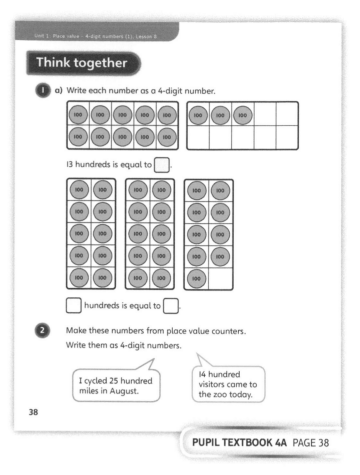

PUPIL TEXTBOOK 4A PAGE 38

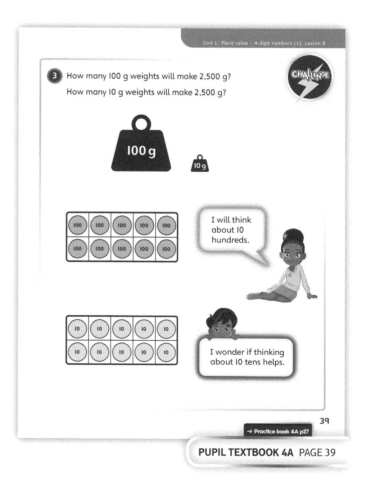

PUPIL TEXTBOOK 4A PAGE 39

Practice

WAYS OF WORKING Independent thinking

IN FOCUS
In question ❶ children exchange 10 hundreds for 1 thousand and in question ❷ they write a multiple of 100 as a 4-digit number. Question ❸ is context-based questions involving multiples of 100 and measure.

STRENGTHEN Ensure that children initially focus on numbers between 1,000 and 2,000, as that is where this language of 100s is more commonly used.

DEEPEN Challenge children to represent given years from a history timeline as 4-digit numbers using place value equipment.

ASSESSMENT CHECKPOINT Question ❶ will allow you to assess whether children can exchange 10 hundreds for 1 thousand in order to write multiples of 100 as 4-digit numbers, and vice versa.

ANSWERS Answers for the **Practice** part of the lesson can be found in the *Power Maths* online subscription.

Reflect

WAYS OF WORKING Independent thinking

IN FOCUS The **Reflect** part of the lesson requires children to explain the conversion of 10s to 100s. Discuss any potential confusion that they encounter around understanding 2,000 as 20 hundreds.

ASSESSMENT CHECKPOINT Assess whether children can justify their explanation of why 20 hundreds are equal to 2,000.

ANSWERS Answers for the **Reflect** part of the lesson can be found in the *Power Maths* online subscription.

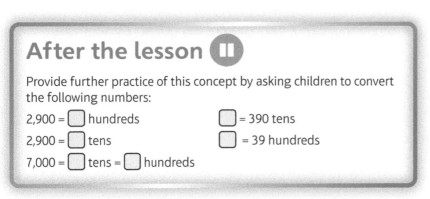

After the lesson ⏸

Provide further practice of this concept by asking children to convert the following numbers:

2,900 = ☐ hundreds   ☐ = 390 tens

2,900 = ☐ tens   ☐ = 39 hundreds

7,000 = ☐ tens = ☐ hundreds

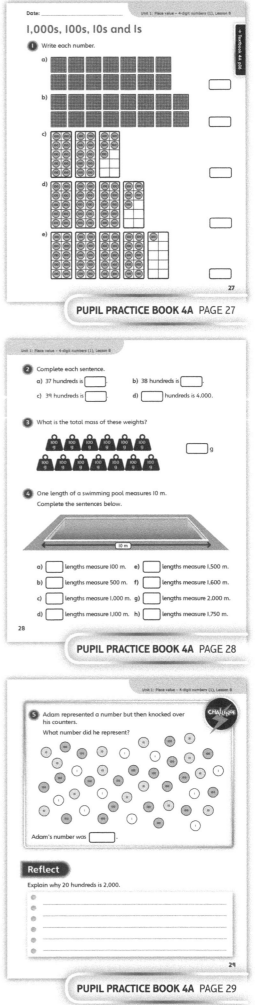

PUPIL PRACTICE BOOK 4A PAGE 27

PUPIL PRACTICE BOOK 4A PAGE 28

PUPIL PRACTICE BOOK 4A PAGE 29

End of unit check

Don't forget the unit assessment grid in your *Power Maths* online subscription.

Group work adult led

IN FOCUS

- Questions **1** to **5** check that children can identify a 4-digit number using different representations.
- In questions **3** and **5** children demonstrate that they know how a number can be partitioned in different ways.
- Question **6** is a SATs-style question and shows understanding of the number line.

ANSWERS AND COMMENTARY Children who have mastered this unit are able to represent 4-digit numbers using a variety of concrete apparatus. They can also use this to find 1, 10, 100, 1,000 more or less than a given number. As well as this, children are able to confidently use a number line up to 1,000.

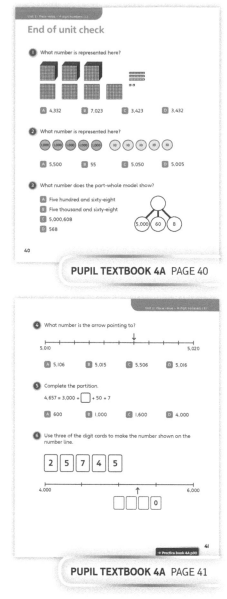

PUPIL TEXTBOOK 4A PAGE 40

PUPIL TEXTBOOK 4A PAGE 41

Q	A	WRONG ANSWERS AND MISCONCEPTIONS	STRENGTHENING UNDERSTANDING
1	D	Choosing A suggests that children have confused the 1,000s and the 100s representation.	Allow children to build the number using base 10 equipment.
2	C	A suggests that children have mistakenly thought the yellow counters were worth 100 rather than 10.	Allow children to place the number on a number line or place value grid to see how to partition 5,050.
3	B	A and D suggest children are not remembering that a place value digit may be 0. They have written the non-zero digits sequentially without inserting 0 for the empty place value.	Provide children with a blank place value grid so they can see where each component belongs.
4	D	A and C suggest children have only looked at the final digit and counted six intervals. B suggests miscounting.	Support children here by counting the increments aloud to help them check their answers.
5	C	A suggests children have looked only at the hundreds digit.	Help children to see that 4,000 – 3,000 = 1,000, so there are 10 hundreds plus another 6 hundreds required.
6	5,250	Children who do not start with 5 have not spotted that 5,000 is the midpoint. Other mistakes indicate they have not found half-way between 5,000 and 5,500.	Support children by counting the increments aloud with them.

My journal

WAYS OF WORKING Independent thinking

ANSWERS AND COMMENTARY

The number shown is 4,563. Children will describe the number using key terms from the unit, including 1,000s, 100s, 10s and 1s.

If children are finding it difficult to put the digits in the correct order, ask: *What number does each piece of base 10 equipment represent? How many 1,000s, 100s, 10s and 1s are there? Which of these comes first?* Encourage them to use a place value grid and put base 10 equipment in the correct columns.

Power check

WAYS OF WORKING Independent thinking

ASK

- *How do you feel about place value in 4-digit numbers?*
- *How confident do you feel showing, describing and partitioning 4-digit numbers in different ways?*
- *How confident do you feel finding 1, 10, 100 or 1,000 more or less than a given number?*
- *What do you need to improve?*

Power play

WAYS OF WORKING Pair work or small groups

IN FOCUS Use this activity to assess whether children can develop a strategy to find out how many 4-digit numbers they can make. Ask what difference it makes to use 6 or 7 counters.

ANSWERS AND COMMENTARY There are many possible answers for this question:

For example, here are the 10 answers that do not use a 0 digit: 1,113, 1,131, 1,311, 3,111, 1,122, 1,212, 1,221, 2,112, 2,121, 2,211.

6 of these round down to 1,000 (also, 3 of them round down to 2,000, and 1 rounds down to 3,000).

If you add an extra counter (7) there are twice as many combinations (20).

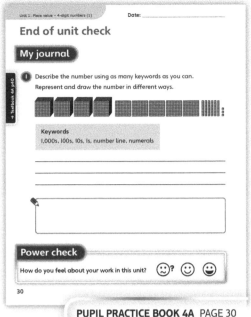

PUPIL PRACTICE BOOK 4A PAGE 30

PUPIL PRACTICE BOOK 4A PAGE 31

After the unit ⏸

- How confident are children with 4-digit numbers?
- Can children partition 4-digit numbers confidently?

Strengthen and **Deepen** activities for this unit can be found in the *Power Maths* online subscription.

Unit 2
Place value – 4-digit numbers ②

Mastery Expert tip! 'Children's understanding of 4-digit numbers will underpin the majority of the work for the year. It is therefore important that children are confident in representing 4-digit numbers in a variety of ways, including using base 10 equipment, place value counters and number lines.'

Don't forget to watch the Unit 2 video!

WHY THIS UNIT IS IMPORTANT

Consolidating learning from the previous unit, children recap that 4-digit numbers are made up of 1,000s, 100s, 10s and 1s; they will be able to represent these numbers in many different ways. Children will look at 4-digit numbers on a number line up to 10,000 and use their understanding of the number line to help compare and order numbers as well as round numbers to the nearest 10, 100 and 1,000. Children will learn to confidently count on and back on number lines. The skills and knowledge that children develop in this unit are fundamental to the rest of their learning within Year 4.

WHERE THIS UNIT FITS

→ Unit 1: Place value – 4-digit numbers (1)
→ **Unit 2: Place value – 4-digit numbers (2)**
→ Unit 3: Addition and subtraction

This unit builds on the previous unit, which introduced 4-digit numbers, emphasising the importance of place value. In the previous unit, children learnt to represent 4-digit numbers and count in 1,000s. In this unit, they will move on to comparing 4-digit numbers and ordering numbers to 10,000. This prepares them for tackling addition and subtraction of 4-digit numbers in the next unit, including numbers where exchanges are needed in more than one column.

Before they start this unit, it is expected that children:

• know that a 4-digit number is made up of 1,000s, 100s, 10s and 1s
• can represent 4-digit numbers in different ways, such as with base 10 equipment, place value grids and counters and part-whole models.

ASSESSING MASTERY

Children can confidently use a number line to order and compare a variety of 4-digit numbers and are able to round to the nearest 10, 100 and 1,000. They will be able to compare and order 4-digit numbers by looking at the digits in each place value column. Children will understand the number line to 10,000 and start to know where numbers lie on the number line. They will be able to round numbers to the nearest 10, 100 and 1,000.

COMMON MISCONCEPTIONS	STRENGTHENING UNDERSTANDING	GOING DEEPER
Children may incorrectly round numbers (to the incorrect 10, 100 or 1,000).	Ensure children have access to a number line so that they can see which 10, 100 or 1,000 their number is closest to and round to that.	Children can try to find any patterns or rules to use when rounding.
Children may compare a 4-digit and a 3-digit number by looking at the first digit rather than the value of the digit.	Use base 10 equipment or place value counters in a place value grid to emphasise that you must compare numbers that have the same place value.	Challenge children to make as many numbers as possible using six counters in a ThHTO place value grid. Ask them to order their numbers on a number line.

UNIT STARTER PAGES

Use these pages to introduce the unit focus to children. You can use the characters to explore different ways of working too!

STRUCTURES AND REPRESENTATIONS

Place value grid, including using base 10 equipment, place value counters and blank counters: This model will help children organise 4-digit numbers into 1,000s, 100s, 10s and 1s, with both concrete representations and abstract numbers.

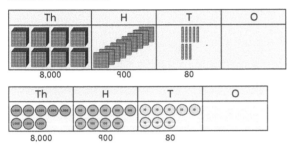

Th	H	T	O
8,000	900	80	

Th	H	T	O
8,000	900	80	

Number line to 10,000: This model will help children to visualise the order of numbers, and can help them to compare numbers. It can also help children to round numbers to the nearest 10, 100 and 1,000.

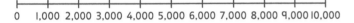

0   1,000 2,000 3,000 4,000 5,000 6,000 7,000 8,000 9,000 10,000

KEY LANGUAGE

There is some key language that children will need to know as part of the learning in this unit.

→ thousands (1,000s), hundreds (100s), tens (10s), ones (1s)
→ place value
→ more, less
→ greater than (>), less than (<), equal to (=)
→ order, compare
→ round up, round down
→ step
→ ascending, descending
→ multiple

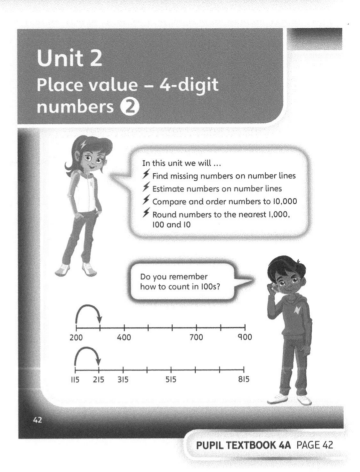

Unit 2
Place value – 4-digit numbers ②

In this unit we will …
⚡ Find missing numbers on number lines
⚡ Estimate numbers on number lines
⚡ Compare and order numbers to 10,000
⚡ Round numbers to the nearest 1,000, 100 and 10

Do you remember how to count in 100s?

200   400   700   900

115 215 315   515   815

42

PUPIL TEXTBOOK 4A PAGE 42

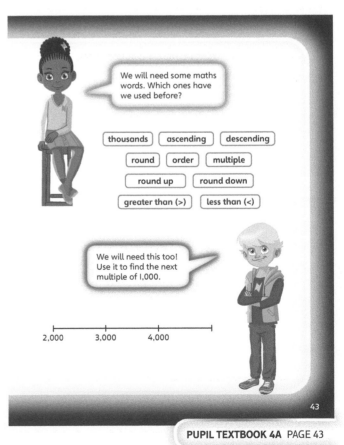

We will need some maths words. Which ones have we used before?

thousands   ascending   descending
round   order   multiple
round up   round down
greater than (>)   less than (<)

We will need this too! Use it to find the next multiple of 1,000.

2,000   3,000   4,000

43

PUPIL TEXTBOOK 4A PAGE 43

Number line to 10,000

Learning focus

In this lesson, children learn to locate and identify multiples of 1,000, 100 and 10 on number lines.

Before you teach

- Do children understand that 4-digit numbers are made from 1,000s, 100s, 10s and 1s?
- Are children able to count on and back in 1,000s, 100s and 10s?

NATIONAL CURRICULUM LINKS

Year 4 Number – number and place value

Identify, represent and estimate numbers using different representations.

Recognise the place value of each digit in a four-digit number (1,000s, 100s, 10s, and 1s).

ASSESSING MASTERY

Children can identify a given 4-digit number based on its location on a number line that has intervals clearly marked.

COMMON MISCONCEPTIONS

Children may not recognise variations in intervals that could show either 10s, 100s or 1,000s. Ask:
- *What can you see on this number line?*
- *Where does this number line start?*
- *What size could each jump be?*

STRENGTHENING UNDERSTANDING

Encourage children to use large number lines. They follow the jumps with finger movements, counting in intervals as their finger moves from left to right or right to left.

GOING DEEPER

Challenge children to explore and describe efficient ways of identifying given numbers on a number line, for example, by counting back from a given number, rather than relying on counting on from left to right every time.

KEY LANGUAGE

In lesson: interval, multiple

STRUCTURES AND REPRESENTATIONS

Number lines

RESOURCES

Mandatory: number lines

Optional: scales

In the eTextbook of this lesson, you will find interactive links to a selection of teaching tools.

Quick recap 🔁

Draw a large number line from 0 to 1,000 with intervals for the 100s.

Count on and back along the line together in 100s.

Discover

ASK

- Question ❶: *What do you notice about this number line?*
- Question ❶ a): *Can you find the half-way point?*
- Question ❶ b): *What sort of counting could help you?*

IN FOCUS Children are identifying all the multiples of 1,000 and then all the multiples of 100 on given number lines. They then locate specific 4-digit numbers on each number line.

PRACTICAL TIPS Act out this activity using a long skipping rope or piece of string. Peg the numbers onto the string and then ask children to check them all for sense.

You could also draw a long line on the board with numbers on sticky notes, or make a line on the playground with whiteboards for the numbers.

ANSWERS

Question ❶ a):

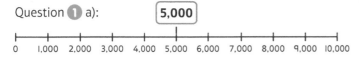

Question ❶ b):

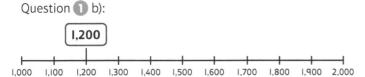

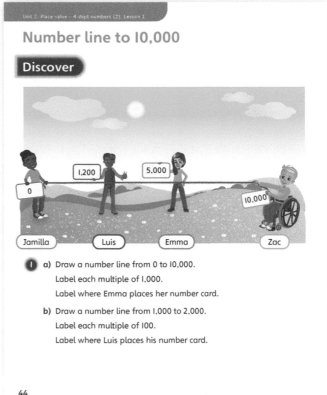

a) Draw a number line from 0 to 10,000.
Label each multiple of 1,000.
Label where Emma places her number card.

b) Draw a number line from 1,000 to 2,000.
Label each multiple of 100.
Label where Luis places his number card.

44

PUPIL TEXTBOOK 4A PAGE 44

Share

ASK

- Question ❶ a): *What size is each jump? Can you count on and back along this number line?*
- Question ❶ b): *What is the magnifying glass for? How does it help us think about the different parts of the line?*

IN FOCUS Question ❶ a) requires children to recognise and locate the multiples of 1,000 between 0 and 10,000.

In question ❶ b), children are locating the multiples of 100 that come between two multiples of 1,000.

Support them in recognising that the second number line is derived from information that is shown on the first number line.

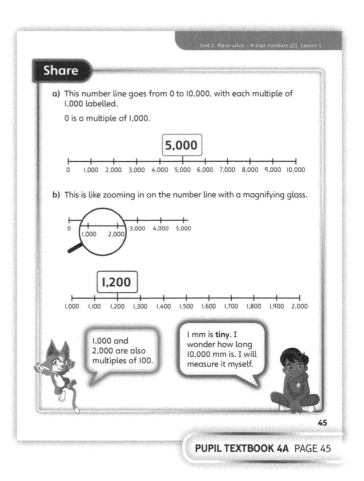

PUPIL TEXTBOOK 4A PAGE 45

Think together

WAYS OF WORKING Whole class teacher led (I do, We do, You do)

ASK

- Question **1**: *What is the size of each jump?*
- Question **2**: *Can you choose whether to count up or down?*
- Question **3**: *What do you notice about the place value of the numbers?*

IN FOCUS In question **1**, children are identifying the multiples of 10 within two 4-digit multiples of 100. Question **2** requires them to identify 4-digit numbers based on their position on two number lines that each show a different scale. In question **2** a) the line is marked in multiples of 10, in question **2** b) it is marked in multiples of 1. Question **3** explores the relationship between 10s, 100s and 1,000s.

STRENGTHEN Provide number lines with numbers written on sticky notes that can be moved and changed depending on which intervals and multiples are being shown.

DEEPEN Challenge children to explore rules for recognising the shared properties of multiples of 10, 100 and 1,000 in more depth. They can use sorting circles to sort multiples of 10, multiples of 100 and multiples of 1,000.

ASSESSMENT CHECKPOINT Question **2** assesses whether children can recognise the intervals shown on a given number line and use this to efficiently identify numbers based on their position on the number line.

ANSWERS

Question **1**: 3,500, 3,510, **3,520**, **3,530**, **3,540**, **3,550**, 3,560, **3,570**, 3,580, **3,590**, **3,600**

Question **2** a): A is 2,490, B is 2,540, C is 2,610

Question **2** b): D is 2,559, E is 2,561, F is 2,569

Question **3**: Danny: 2,000, 10,000, 5,000
Reena: 4,500, 2,000, 10,000, 1,200, 2,100, 3,900, 5,000
Max: 4,500, 2,000, 3,290, 10,000, 9,990, 1,200, 2,100, 2,010, 3,900, 5,000, 1,010
All three children: 2,000, 10,000, 5,000

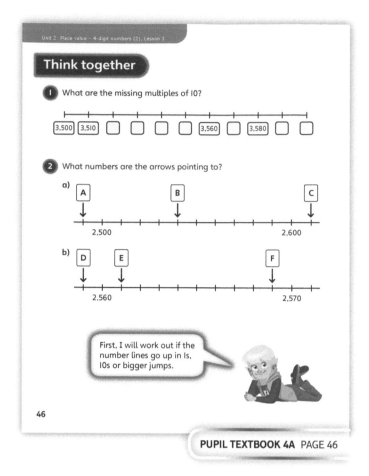

PUPIL TEXTBOOK 4A PAGE 46

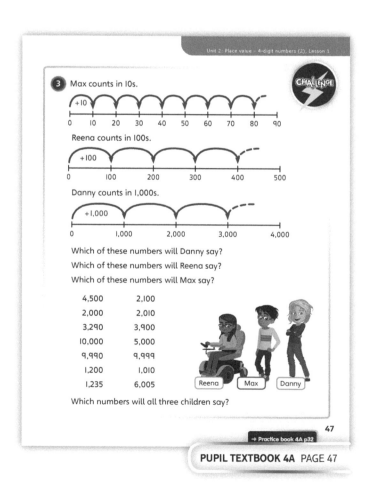

PUPIL TEXTBOOK 4A PAGE 47

Practice

WAYS OF WORKING Independent thinking

IN FOCUS In question ❶, children identify numbers from their position indicated on a given number line. Question ❷ requires children to continue counts on number lines in multiples of 1, 10 or 100. In order to complete question ❸, children will need to identify what the intervals are on each number line. In questions ❹ and ❻, children apply their knowledge of multiples in the context of measure. In question ❺, children should notice that the number lines both bridge multiples of 1,000. Do they understand what this will mean for the missing numbers they are trying to find?

STRENGTHEN Give children opportunities to use measuring equipment with scales to explore number lines in a meaningful practical context.

DEEPEN Ask children to list multiples of 10, 100 or 1,000 that they have encountered in a range of contexts, such as mass, length or capacity.

ASSESSMENT CHECKPOINT Question ❸ assesses whether children can identify the intervals used on several number lines that each show a different scale.

ANSWERS Answers for the **Practice** part of the lesson can be found in the *Power Maths* online subscription.

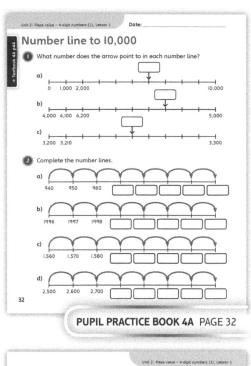

PUPIL PRACTICE BOOK 4A PAGE 32

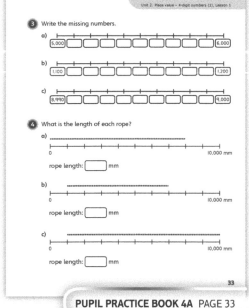

PUPIL PRACTICE BOOK 4A PAGE 33

Reflect

WAYS OF WORKING Pair work

IN FOCUS The **Reflect** part of the lesson provides the opportunity for children to discuss a potential misconception when counting in regular intervals between two multiples of 100. The student in the question has misinterpreted the scale of the line and counted up in 1s, rather than 10s.

ASSESSMENT CHECKPOINT Assess whether children can explain the mistake and describe how they would recognise and label the correct intervals on this number line.

ANSWERS Answers for the **Reflect** part of the lesson can be found in the *Power Maths* online subscription.

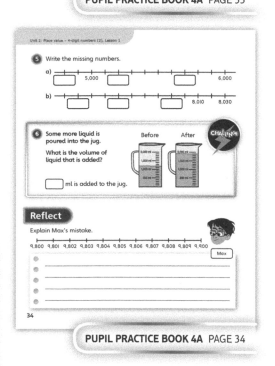

PUPIL PRACTICE BOOK 4A PAGE 34

After the lesson ⏸

- Can children draw a number line that correctly shows multiples of 1,000?
- Are they able to identify multiples of 10 or 100 shown on given number lines?

Between two multiples

Learning focus

In this lesson, children identify numbers in a range between two multiples of 1,000, 100 or 10. They also identify the previous and next multiple of 1,000, 100 or 10 that a given number lies between.

Before you teach

- Are children able to count up in 10s, 100s and 1,000s?
- Can children draw number lines that show multiples of 10, 100 or 1,000?

NATIONAL CURRICULUM LINKS

Year 4 Number – number and place value

Recognise the place value of each digit in a four-digit number (1,000s, 100s, 10s, and 1s).

Count in multiples of 6, 7, 9, 25 and 1,000.

ASSESSING MASTERY

Children can identify the previous and next multiple of 1,000, 100 or 10, that come before and after a given number with up to four digits.

COMMON MISCONCEPTIONS

Children may not recognise that a multiple of 1,000 is also a multiple of 10 and 100. Ask:
- *What do you notice about all these multiples of 10?*
- *Could a multiple of 10 also be a multiple of 100?*
- *Is 100 a multiple of 10? Is 0 a multiple of 10? Of 100? Of 1,000?*

STRENGTHENING UNDERSTANDING

Spend time consolidating multiples of 10, 100 and 1,000 by giving children opportunities to count, as a class, in these multiples.

GOING DEEPER

Ask children to explore different ways to estimate a number on a number line with increasing precision when it is positioned on an open interval.

KEY LANGUAGE

In lesson: multiple, intervals

STRUCTURES AND REPRESENTATIONS

Number lines

RESOURCES

Mandatory: number lines

Optional: sticky notes, pegs

 In the eTextbook of this lesson, you will find interactive links to a selection of teaching tools.

Quick recap

Draw a large number line from 0 to 10,000 with intervals for the 1,000s. Count on and back along the line together in 1,000s.

Discover

Between two multiples

Discover

I have done between 3,000 and 4,000 steps today.

Andy

I have done exactly 6,790 steps so far.

Bella

WAYS OF WORKING Pair work

ASK

- Question ① a): *What number can you find in the picture? Is this a big number or a small number? Why do you say that?*
- Question ① b): *What do you think Bella's number might be? Are there any numbers that it will definitely not be? How can you convince yourself of that?*

IN FOCUS Children first identify the multiples of 1,000 on a number line to 10,000. They then explore the range of numbers that come between two multiples of 1,000.

PRACTICAL TIPS Make a large number line from 0 to 10,000. Ask children to use sticky notes or pegs to place the numbers in approximate positions. They can then move the numbers to make their estimates increasingly accurate.

ANSWERS

Question ① a):

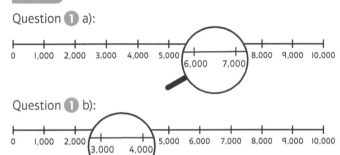

Question ① b):

Bella's number is on this part of the line

① a) Draw a number line from 0 to 10,000.
Label the multiples of 1,000.
Andy's number is between two multiples. Which ones?

b) Show the part of the line for Bella's number.

48

PUPIL TEXTBOOK 4A PAGE 48

Share

WAYS OF WORKING Whole class teacher led

ASK

- Question ① a): *Where does the number line start? How big is each jump?*
- Question ① a): *Can you count up and back on this number line? What is the magnifying glass helping us to think about?*
- Question ① a): *How can you tell that 6,790 is greater than 6,000 but less than 7,000?*
- Question ① b): *Can you name some numbers that are in this interval? What is the same about all the numbers between 3,000 and 4,000?*

IN FOCUS In question ① a), children are using a number line to identify the previous and next multiples of 1,000 for a given 4-digit number.

For question ① b), children locate two consecutive multiples of 1,000 on the number line. They can also consider the range of numbers that will fall between those two multiples of 1,000.

Share

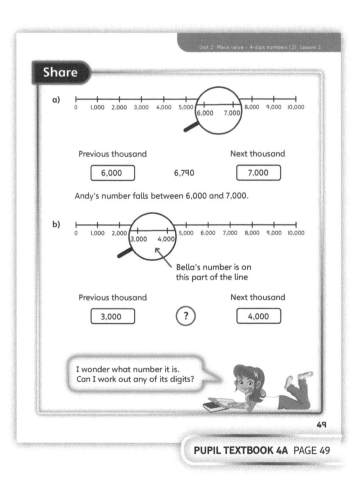

a)

Previous thousand
6,000

6,790

Next thousand
7,000

Andy's number falls between 6,000 and 7,000.

b)

Bella's number is on this part of the line

Previous thousand
3,000

?

Next thousand
4,000

I wonder what number it is. Can I work out any of its digits?

49

PUPIL TEXTBOOK 4A PAGE 49

Think together

ASK

- Question **1**: *Could you start counting from this number? What is the same and what is different about any number in this range? What do you know about the digits?*
- Question **2**: *Which digits help you think about the previous and next multiples of 1,000, 100 or 10?*
- Question **3**: *Could you draw a number line to help you?*

IN FOCUS In question **1**, children are identifying numbers that fall between two given multiples on number lines with different scales. Question **2** requires them to then identify the previous and next multiples of 100 or 10 for given numbers. In question **3**, children examine the same number in different ways to understand the relationship between multiples of 10, 100 and 1,000.

STRENGTHEN To support work in question **1**, make a large number line and write numbers on sticky notes, so that children can explore what happens when the scale of the line changes each time.

DEEPEN In question **3**, children should notice that the next 1,000, 100 and 10 are all the same. Ask them to explain why that is the case. Ask: *What other numbers would this happen for? When would it not happen? Can you think of any numbers where the previous 1,000, 100 and 10 are all the same? Why is this?*

ASSESSMENT CHECKPOINT Question **2** assesses whether children can use their understanding of place value to support their reasoning when finding the previous and next multiple.

ANSWERS

Question **1** a): Any number between 7,001 and 7,999.

Question **1** b): Any number between 2,201 and 2,299.

Question **1** c): Any number between 1,781 and 1,789.

Question **2**:

a): Multiples of 100:

previous	number	next
1,500	1,511	**1,600**
2,700	2,778	**2,800**
3,900	3,964	**4,000**
0	26	**100**
800	889	**900**
2,500	2,501	**2,600**

b): Multiples of 10:

previous	number	next
1,510	1,511	**1,520**
2,770	2,778	**2,780**
3,960	3,964	**3,970**
20	26	**30**
880	889	**890**
2,500	2,501	**2,510**

Question **3** a): 4,000 and 5,000

Question **3** b): 4,900 and 5,000

Question **3** c): 4,990 and 5,000

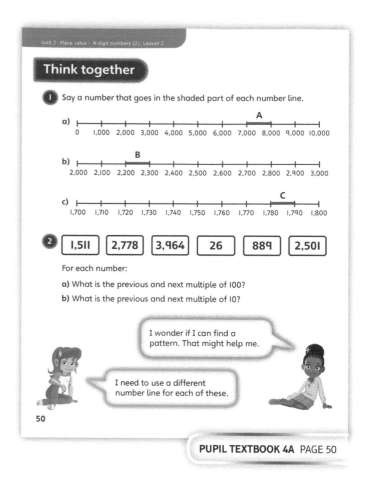

PUPIL TEXTBOOK 4A PAGE 50

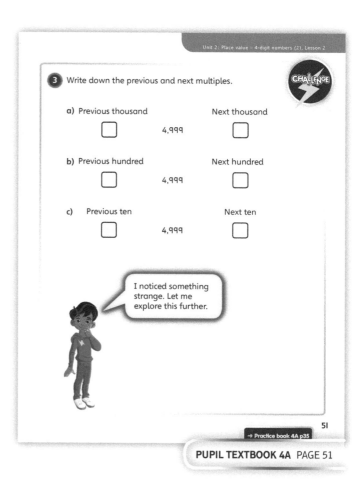

PUPIL TEXTBOOK 4A PAGE 51

Practice

WAYS OF WORKING Independent thinking

IN FOCUS In questions **1**, **2** and **3**, children are identifying numbers that will fall within a given range, between two multiples of 1,000, 100 or 10. The number line supports them in doing this. Make sure children notice that the scale of the number line changes each time.

Question **4** requires children to identify the previous and next multiples of 1,000, 100 or 10 for a given number. The number line is no longer provided, but children may still find it helpful to draw their own.

In question **5**, children solve missing number puzzles by applying their knowledge of place value to find numbers that fall between previous and next multiples of 1,000, 100 or 10.

STRENGTHEN Make a large number line together. Ask children to write the numbers on sticky notes so that the numbers can be changed and moved around depending on which question they are working on.

DEEPEN Ask children to make their own missing number puzzles like the ones in question **5**, using digit cards that can be covered or turned over.

ASSESSMENT CHECKPOINT Question **4** assesses whether children can use their understanding of place value to identify previous and next multiples of 10, 100 and 1,000.

ANSWERS Answers for the **Practice** part of the lesson can be found in the *Power Maths* online subscription.

Reflect

WAYS OF WORKING Independent thinking

IN FOCUS The **Reflect** part of the lesson prepares children for thinking about contexts that will lead to the idea of rounding.

ASSESSMENT CHECKPOINT Assess whether children can suggest plausible contexts for using previous and next multiples. These should focus on finding a suitable level of accuracy or making guesses.

ANSWERS Answers for the **Reflect** part of the lesson can be found in the *Power Maths* online subscription.

After the lesson ⏸

- Play a 'Guess my number' game. Encourage children to make guesses such as: *Is the number between 2,000 and 3,000?* Do they understand what information will help them to make sensible guesses about your number?

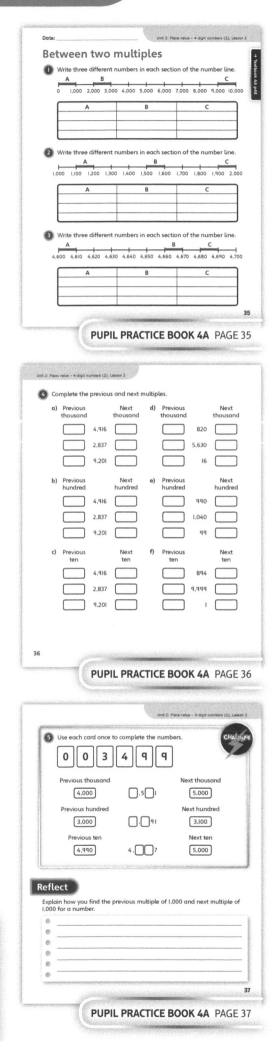

PUPIL PRACTICE BOOK 4A PAGE 35

PUPIL PRACTICE BOOK 4A PAGE 36

PUPIL PRACTICE BOOK 4A PAGE 37

Estimate on a number line to 10,000

Learning focus

In this lesson, children develop their knowledge of place value and comparison of numbers to make sensible estimates on a number line.

NATIONAL CURRICULUM LINKS

Year 4 Number – number and place value

Order and compare numbers beyond 1,000.

Identify, represent and estimate numbers using different representations.

ASSESSING MASTERY

Children can justify their reasoning for making improved estimates about the location of numbers on a number line.

COMMON MISCONCEPTIONS

Children may not be confident in reasoning proportionally about the scale of a number line. Ask:

- *What steps could you count on this number line?*
- *Where does this number line begin and end?*
- *What does each interval represent on this number line?*

STRENGTHENING UNDERSTANDING

Start by working with number lines that are in lower ranges, for example 0 to 10, 0 to 20, 0 to 100, and so on, to build children's confidence.

GOING DEEPER

Challenge children to explore different strategies for improving their estimates, for example, finding half-way points between numbers.

KEY LANGUAGE

In lesson: estimate, half-way

Other language used by the teacher: interval, number line, middle, scale

STRUCTURES AND REPRESENTATIONS

Number line

RESOURCES

Optional: number lines, sticky notes

 In the eTextbook of this lesson, you will find interactive links to a selection of teaching tools.

Quick recap 🔍

Draw a large number line from 0 to 100 and ask children to estimate the position of 25, 50, 67 and 99.

Discover

WAYS OF WORKING Pair work

ASK

- Question ① a): *How would the arrow move along the scale as Sofia runs further?*
- Question ① a): *What are the intervals on this number line?*
- Question ① b): *Can you count on in the intervals from 0?*
- Question ① b): *Can you say a number that is less than the distance shown? Can you say a number that is greater than the distance shown?*

IN FOCUS Children are recognising the intervals on a given number line in the context of measuring distance, and using this to make judgements about numbers on the line.

They will also need to recognise previous and next multiples of 100, which builds on learning from the previous lesson.

PRACTICAL TIPS Count on along a large number line like the one in the question. Provide an arrow that can be moved along the line as you go.

ANSWERS

Question ① a): Sofia has run between 1,000 m and 1,100 m.

Question ① b): A good estimate is 1,070 m.

Estimate on a number line to 10,000

Discover

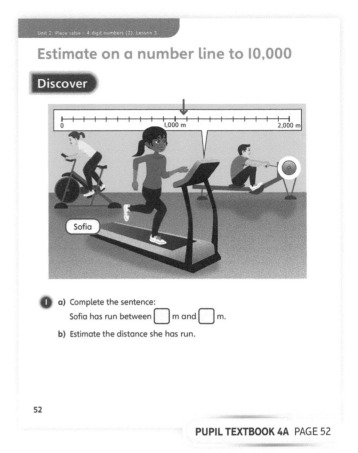

① a) Complete the sentence:
Sofia has run between ☐ m and ☐ m.

b) Estimate the distance she has run.

52

PUPIL TEXTBOOK 4A PAGE 52

Share

WAYS OF WORKING Whole class teacher led

ASK

- Question ① a): *Can you see the previous and next multiples?*
- Question ① a): *How is this the same and how is this different to the last lesson?*
- Question ① b): *Why is the number 1,050 important?*
- Question ① b): *Can you prove that 1,050 is half-way between 1,000 and 1,100?*

IN FOCUS In order to answer question ① a), children will need to identify the intervals on the number line and then use these to find the previous and next multiples of 100 for the number shown. They then use this information to make a sensible estimate in question ① b). They should justify their estimate by reasoning about the nearest multiple of 100 to it.

Share

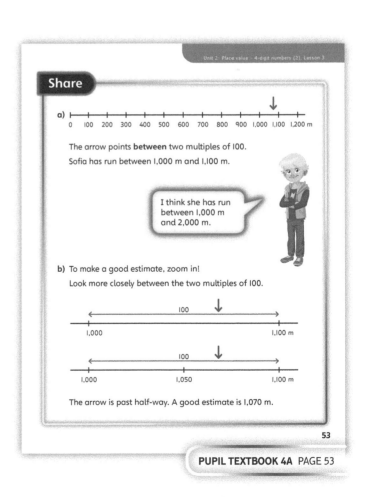

53

PUPIL TEXTBOOK 4A PAGE 53

Think together

Whole class teacher led (I do, We do, You do)

ASK

- Question **1**: *How big is the jump from the start to the end of each line?*
- Question **2**: *Can you make an estimate that is too small? Too big? How close can you get to the arrow?*
- Question **3**: *What is the same and what is different about each number line?*

IN FOCUS Question **1** has three examples where children need to identify the half-way point of one interval on a number line and then use this to find a number within a given range. In question **2** they consolidate this learning by using the information provided on a number line to make a reasoned estimate. Question **3** requires children to reason about possible variations in number lines, one of which does not start at 0.

STRENGTHEN Use large number lines and write numbers on sticky notes so that children can move the numbers in order to refine their estimates.

DEEPEN Challenge children to explore estimates that can be made using a range of scales in the context of measuring equipment, for example, measuring capacity.

ASSESSMENT CHECKPOINT Question **1** assesses whether children can identify and reason about the half-way point in intervals of 1,000, 100 or 10.

ANSWERS

Question **1** a): 5,500; A: any number between 5,001 and 5,499, B: any number between 5,501 and 5,999.

Question **1** b): 6,450; A: any number between 6,401 and 6,449, B: any number between 6,451 and 6,499.

Question **1** c): 2,585; A: 2,581, 2,582, 2583 or 2,584, B: 2,586, 2,587, 2,588 or 2,589.

Question **2**: The number is between 3,000 m and 3,500 m. It is approximately 3,200 m.

Question **3** a):

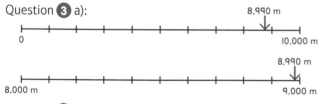

Question **3** b): Answers will vary depending on the starting and ending values of the number lines, and the scale. However, it should be the case that:

3,001 is very close to 3,000

5,900 is close to and before 6,000

7,500 is half-way between 7,000 and 8,000.

Think together

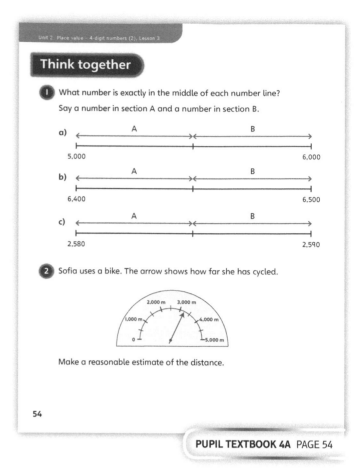

1 What number is exactly in the middle of each number line?

Say a number in section A and a number in section B.

2 Sofia uses a bike. The arrow shows how far she has cycled.

Make a reasonable estimate of the distance.

54

PUPIL TEXTBOOK 4A PAGE 54

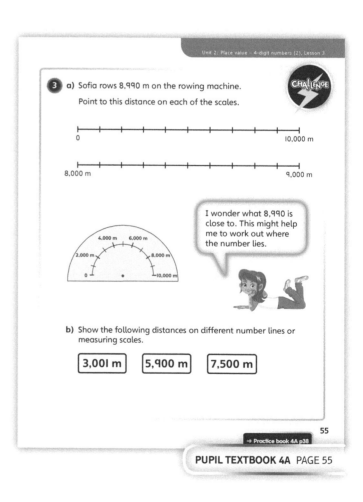

3 a) Sofia rows 8,990 m on the rowing machine.

Point to this distance on each of the scales.

I wonder what 8,990 is close to. This might help me to work out where the number lies.

b) Show the following distances on different number lines or measuring scales.

| 3,001 m | 5,900 m | 7,500 m |

55

→ Practice book 4A p38

PUPIL TEXTBOOK 4A PAGE 55

Practice

WAYS OF WORKING Independent thinking

IN FOCUS Question ❶ asks children to make estimates about numbers that are located on number lines with intervals of 1,000 and then 100. In question ❷, they are estimating the position of given numbers on number lines with intervals of 1,000, 100 or 10. In questions ❸, ❹ and ❺, children are making estimates with scales that are in the context of measuring equipment, including measuring volume and mass. Question ❻ is a problem-solving exercise involving estimation of length.

STRENGTHEN Use large number lines and write numbers on sticky notes so that children can move the numbers in order to refine their estimates.

DEEPEN Give children the opportunity to practically explore making estimates of length in millimetres that are between 0 and 10,000 mm (0 and 10 m).

ASSESSMENT CHECKPOINT Question ❷ assesses whether children can make reasonable estimates on number lines that represent different intervals.

ANSWERS Answers for the **Practice** part of the lesson can be found in the *Power Maths* online subscription.

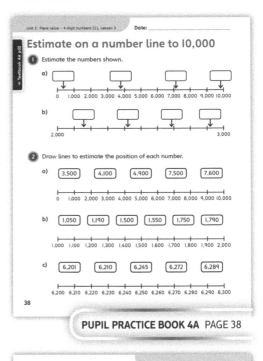

PUPIL PRACTICE BOOK 4A PAGE 38

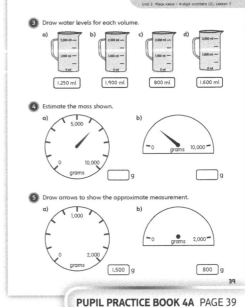

PUPIL PRACTICE BOOK 4A PAGE 39

Reflect

WAYS OF WORKING Independent thinking

IN FOCUS The **Reflect** part of the lesson suggests a game where children will apply their estimation skills to place numbers on a 0 to 10,000 number line.

ASSESSMENT CHECKPOINT Assess whether children can position numbers on the line with reasonable accuracy. Are they able to make estimates that meet the criteria specified in the question?

ANSWERS Answers for the **Reflect** part of the lesson can be found in the *Power Maths* online subscription.

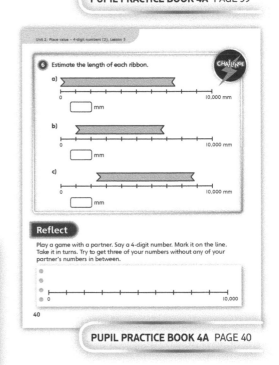

PUPIL PRACTICE BOOK 4A PAGE 40

After the lesson ⏸

- Can children draw a 0 to 10,000 number line and then place a selection of numbers on it with reasonable accuracy?
- What other opportunities can you find for children to explore why making estimates is a useful and important maths skill?

Compare and order numbers to 10,000

Learning focus

In this lesson children will order 4-digit numbers, focusing on the value of the digits and using a place value grid to support understanding.

Before you teach [II]

- Can children partition a 4-digit number?
- Can children use the <, > and = signs?
- Can children compare two 4-digit numbers?

NATIONAL CURRICULUM LINKS

Year 4 Number – number and place value

Order and compare numbers beyond 1,000.

Identify, represent and estimate numbers using different representations.

ASSESSING MASTERY

Children can order numbers by focusing on the values of the digits, using a place value grid to support them. Children can confidently use language such as 'greatest', 'smallest', 'descending' and 'ascending' to order numbers efficiently.

COMMON MISCONCEPTIONS

Children may need reminding about the order in which we compare numbers. To prompt their thinking, ask:
- *Which digits should you compare first? Why?*
- *Are the first digits of each number of the same value? How do you know?*
- *If the 1,000s are the same, what do you look at next?*

STRENGTHENING UNDERSTANDING

Encourage children to use concrete equipment such as base 10 equipment to build the numbers they are working with. This will show the distinctions between the values of each digit in a number. Placing the concrete equipment into a place value grid, and saying out loud what each digit is worth, will further reinforce understanding of place value and order. Children can also use a number line to help them visualise the order of numbers. They should observe that the greater a number is, the further along the number line it will be.

GOING DEEPER

To prompt deep engagement in this lesson, encourage children to complete the missing digit questions by reasoning about the order of the numbers. Challenge children by asking if they can exhaust all possible solutions in each question.

KEY LANGUAGE

In lesson: fewest, least, order, smallest, greatest, most, thousands (1,000s), hundreds (100s), tens (10s), ones (1s), highest, lowest, ascending, descending

Other language to be used by teacher: furthest, shortest

STRUCTURES AND REPRESENTATIONS

Place value grids, number lines

RESOURCES

Mandatory: place value grids

Optional: base 10 equipment, place value counters, number lines

 In the eTextbook of this lesson, you will find interactive links to a selection of teaching tools.

Quick recap 🔄

Ask children to find the smallest and the greatest number from this set:

201, 99, 450, 1,003, 100, 999.

Discuss how they can justify their choices.

Discover

Pair work

ASK

- Question ① a): *What are the children's scores? If you are looking for who scored the fewest points, are you looking for the greatest or smallest number?*
- Question ① b): *Who has the most points? Is the greatest number 1st? Is this always true?*

IN FOCUS Question ① a) encourages children to think about the link between the least/fewest of something being the smallest number of something. Children will build on their learning from the previous lesson by comparing digits abstractly.

PRACTICAL TIPS Children can make the numbers using place value counters or plain counters on a place value grid to help them order the numbers more efficiently. Encourage children to present the numbers one above another on the same grid to aid direct comparison. It is vital that children are confident with the place values of 4-digit numbers and comparing two numbers before moving on to ordering numbers.

ANSWERS

Question ① a): Mo has scored the fewest points.

Question ① b): 8,645 > 8,632 > 8,052

Jamie came 1st. Olivia came 2nd.
Amelia came 3rd.

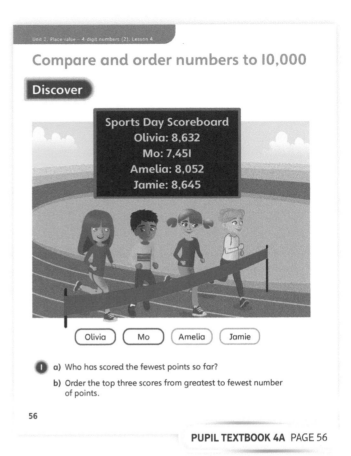

PUPIL TEXTBOOK 4A PAGE 56

Share

WAYS OF WORKING Whole class teacher led

ASK

Question ① a): *Does putting the numbers in the same grid help you find the smallest number? How did you decide which is the smallest number?*

Question ① b): *How can you find the greatest number? What do you need to compare first? What challenge do you notice? What do you have to do next?*

IN FOCUS Question ① a) has been designed to show how presenting numbers in the same table makes comparison easier, and that comparing the 1,000s first is not only the correct way, but also the most efficient way to compare these four numbers. One number should stand out as being smaller than the others. Ensure that children are using the key vocabulary of place value, and that you model the use of key comparison vocabulary such as 'smallest', 'fewest', 'least', 'less than', 'greatest', 'most', 'more', 'more than', and so on.

STRENGTHEN Displaying each number using counters or base 10 equipment to help children see why one number is greater than or less than another number will help to secure understanding in this lesson.

DEEPEN Can children use the comparison signs >, < and = to compare the numbers in question ① b)?

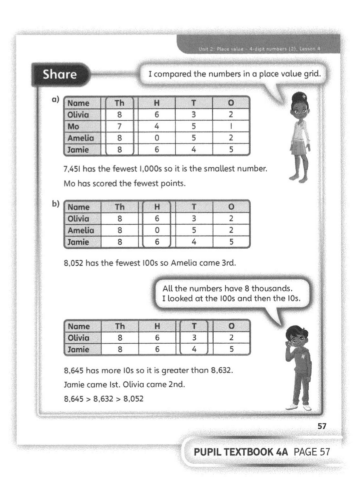

PUPIL TEXTBOOK 4A PAGE 57

Think together

WAYS OF WORKING Whole class teacher led (I do, We do, You do)

ASK

- Question **1**: *Which digit do you look at first? Why?*
- Question **2**: *If you are writing the numbers in descending order, will the smallest or greatest number be at the beginning? What is the opposite of descending?*
- Question **3**: *Will putting the numbers on a number line help you to compare them? How?*
- Question **3**: *What word do you use to describe the order that you have put the masses in?*

IN FOCUS Questions **1** and **2** build on learning from the **Discover** task, requiring children to practise the principles they applied in **Discover**. Check that children are secure using the signs of comparison. Can children answer the questions without using a place value grid to support them? In question **2**, encourage children to repeat out loud the new vocabulary 'descending' and 'ascending'.

Question **3** challenges children to use a number line to compare and order numbers that do not all have the same number of digits. Refer children to what Astrid and Dexter are saying to give them some pointers if necessary.

STRENGTHEN Reinforce understanding of effective strategies for ordering numbers by reading the questions aloud. Read aloud as a class each number to be compared and model comparing the 1,000s first, then the 100s, and so on. This activity will help children build the procedural memory that is required to accurately order large numbers.

DEEPEN Challenge children by setting problems where they have to reason what a missing digit might be, based on other numbers in a sequence. For example: 3,☐26 < ☐,52☐ < 3,5☐8. Ask: *How many different answers can you find? What sort of reasoning are you using to help you find the missing digits? If the first missing digit is a 5, how does this limit what the other missing digits can be?*

ASSESSMENT CHECKPOINT In question **2** check children understand the new language of 'ascending' and 'descending'.

ANSWERS

Question **1** a): 8,624 > 8,426

Question **2** a): 240 < 1,028 < 1,220 < 1,250

Question **2** b): 1,250 > 1,220 > 1,028 > 240.

Question **3**: Children may suggest using a number line to compare the masses, or looking at the 1,000s digit first, then the 100s and so on.

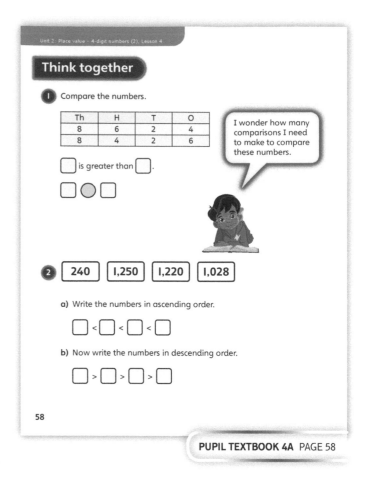

PUPIL TEXTBOOK 4A PAGE 58

PUPIL TEXTBOOK 4A PAGE 59

Practice

WAYS OF WORKING Independent thinking

IN FOCUS Question ① requires children to practise the main principles of the lesson and compare the numbers.

Question ② revisits learning from earlier lessons, testing children's ability to use the < and > symbols for comparison. In question ④, children move from comparing to ordering, with the place value grid for support.

In question ⑤ children need to think about whether they should order numbers differently if a unit of measurement such as kg or m is included with the number. Ask: *Does this affect the order?*

Question ⑥ consolidates learning of key comparison and number ordering vocabulary, including 'furthest', 'second' and 'shortest'.

In question ⑦ children must reason what the missing numbers are. It is important that children do not just look at the next number in the list, but instead look at all the numbers to help them make informed decisions about what the missing numbers could be (and therefore what they cannot be).

STRENGTHEN To support children in working out answers to the missing digit statements in question ③, suggest that they first use a trial and error approach, guessing a number and then checking if their guessed number works. Encourage discussion around the numbers they have guessed to support their reasoning about whether their guess works or not. If it doesn't work, why not? Take opportunities to say the numbers out loud with children, saying, for example, *5 thousands are greater than 4 thousands.*

DEEPEN Explore children's thinking in question ⑦. What strategies do they use to find out what the missing numbers might be? How do they work out which 4-digit numbers have digit sums of 15 and then apply this to answering the question?

ASSESSMENT CHECKPOINT Assess whether children can order numbers in ascending and descending order. Are children able to find missing digits in comparison number sentences, and to use ordering to find missing numbers?

ANSWERS Answers for the **Practice** part of the lesson can be found in the *Power Maths* online subscription.

Reflect

WAYS OF WORKING Independent thinking

IN FOCUS Children use their knowledge of 4-digit numbers to make and then compare and order numbers with the same digits in different orders. Look for children who can order their numbers quickly and those that need additional support.

ASSESSMENT CHECKPOINT Assess whether children can explain how they ordered their numbers.

ANSWERS Answers for the **Reflect** part of the lesson can be found in the *Power Maths* online subscription.

After the lesson

- Do children rely on the place value grid to help them order numbers?
- How confident are children at reasoning to help them work out missing digits?
- Can children explain how they know that they have ordered them correctly?

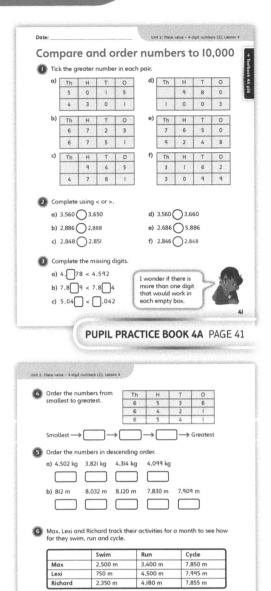

PUPIL PRACTICE BOOK 4A PAGE 41

PUPIL PRACTICE BOOK 4A PAGE 42

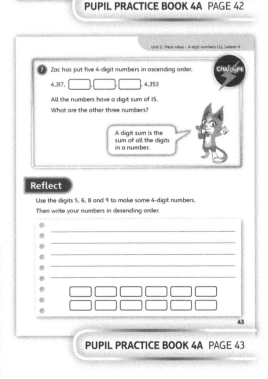

PUPIL PRACTICE BOOK 4A PAGE 43

Round to the nearest 1,000

Learning focus

In this lesson, children round 4-digit numbers to the nearest 1,000, building on their work with finding previous and next multiples of 10, 100 and 1,000.

Before you teach

- Are there any additional misconceptions you need to consider?
- Do children know the place value of each digit in a 4-digit number?

NATIONAL CURRICULUM LINKS

Year 4 Number – number and place value

Round any number to the nearest 10, 100 or 1,000.

ASSESSING MASTERY

Children can justify their rounding based on their understanding of the closest multiple of 1,000 to a given number.

COMMON MISCONCEPTIONS

Children may not see rounding as being as accurate or as good as finding the 'exact' number. Ask:
- *What is the previous multiple? What is the next multiple? Which is closest?*
- *When might it be useful to know this?*

STRENGTHENING UNDERSTANDING

Focus on asking children to find the 'closest' multiple of 1,000, and introduce the language structure clearly. For example:

3,402 is between 3,000 and 4,000. 3,402 is closer to 3,000 than to 4,000.

3,402 rounds to 3,000. The nearest 1,000 is 3,000.

3,402 rounds to 3,000 to the nearest 1,000.

GOING DEEPER

Ask children to give examples of when they might need to round to the nearest 1,000. How many examples can they think of?

KEY LANGUAGE

In lesson: round, **rounds up**, **rounds down**, nearest, between, closer, closest, multiple

Other language to be used by the teacher: difference, half-way

STRUCTURES AND REPRESENTATIONS

Number lines

RESOURCES

Mandatory: number lines

 In the eTextbook of this lesson, you will find interactive links to a selection of teaching tools.

Quick recap

Play 'Guess my number'. Say: *I am thinking of a number between 100 and 200.*

Ask children to suggest possible answers. You can respond with *Your guess is too high* or *Your guess is too low.*

Discover

WAYS OF WORKING Pair work

ASK

- Question ① a): *Can you say the 4-digit number to a partner?*
- Question ① a): *Could you estimate its position on a number line?*
- Question ① b): *Which digits in this number would help you decide where it sits on a number line?*

IN FOCUS Children are finding the previous and next multiples of 1,000 for a given 4-digit number, in the context of money. They are then identifying which of these is the closest multiple of 1,000. This begins to introduce the concept of rounding.

PRACTICAL TIPS Allow children to use concrete equipment to represent the numbers, helping to support their decision about whether they need to round up or down.

ANSWERS

Question ① a): 5,275 is between 5,000 and 6,000.

Question ① b): 5,275 is closer to 5,000 than 6,000.

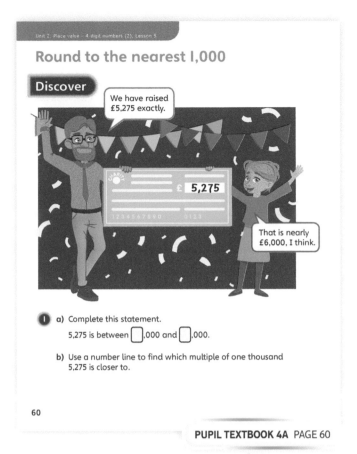

PUPIL TEXTBOOK 4A PAGE 60

Share

WAYS OF WORKING Whole class teacher led

ASK

- Question ① a): *How can you justify that 5,275 is between 5,000 and 6,000?*
- Question ① a): *What is it about the number 5,275 that tells you it is greater than 5,000 but less than 6,000?*
- Question ① a): *What could you say to prove that 5,500 is exactly half-way between 5,000 and 6,000?*
- Question ① b): *How do you know 5,275 is less than 5,500?*

IN FOCUS Question ① b) addresses rounding. When you are rounding to the nearest 1,000 and looking at a number line, the distance from the number to each end of the number line can help you to decide which number to choose. This also links to children's previous knowledge of place value.

STRENGTHEN Provide a large number line that goes from 5,000 to 6,000. Count together up the line in intervals of 100.

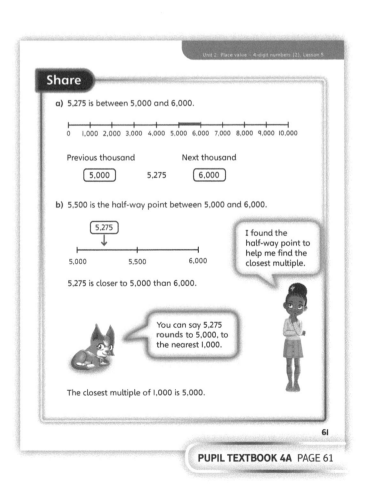

PUPIL TEXTBOOK 4A PAGE 61

Think together

WAYS OF WORKING Whole class teacher led (I do, We do, You do)

ASK

- Questions ❶: *When you read the number 2,891, what tells you where to look for it on the number line?*
- Questions ❷: *Is 0 a multiple of 1,000?*
- Questions ❸: *What decisions do you make when thinking about the nearest 1,000?*

IN FOCUS Question ❶ provides opportunities for children to practise rounding numbers. Encourage children to work out the 1,000 on either side of the number first, and then to work out where to place this on a number line.

Question ❷ focuses on rounding without the visual aid of the number line. Children will now need to look at the 100s digit in order to round this number accurately.

Question ❸ develops children's understanding of place value in relation to rounding.

STRENGTHEN Encourage children to draw their own number lines that show the previous and next multiples of 1,000 that each number sits between.

DEEPEN Ask children to investigate how to use the place value of the 100s digit in order to decide on the correct rounding.

ASSESSMENT CHECKPOINT Question ❷ assesses whether children can round accurately to the nearest 1,000.

ANSWERS

Question ❶: 2,891 > 2,850 so rounds to 3,000.

Question ❷ a): Ends of the number line are 6,000 and 7,000. 6,200 rounds to 6,000 to the nearest thousand.

Question ❷ b): Ends of the number line are 3,000 and 4,000. 3,760 rounds to 4,000 to the nearest thousand

Question ❷ c): Ends of the number line are 0 and 1,000. 862 rounds to 1,000 to the nearest thousand.

Question ❸: 2,470 rounds down to 2,000.
2,883 rounds up to 3,000.
7,500 rounds up to 8,000.
3,782 rounds up to 4,000.
9,501 rounds up to 10,000.

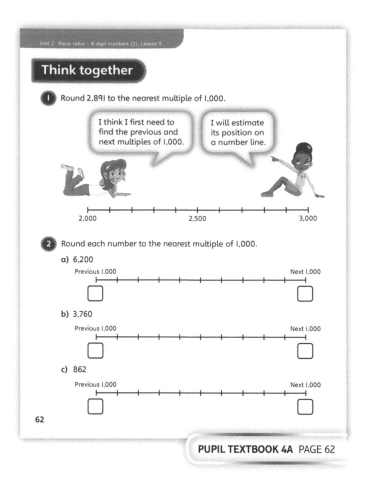

PUPIL TEXTBOOK 4A PAGE 62

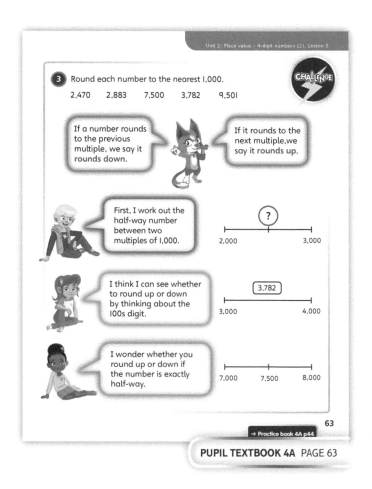

PUPIL TEXTBOOK 4A PAGE 63

Practice

WAYS OF WORKING Independent thinking

IN FOCUS In question ❶ children are rounding using a number line which is provided. Question ❷ requires children to choose the nearest multiple of 1,000. They then apply this learning to question ❸ where they round a given 4-digit number to the nearest 1,000. In question ❹, encourage children to look for some numbers that will round up and some that will round down to the nearest 1,000.

Question ❺ requires children to think about which values are important when rounding to the nearest 1,000. Ask: *Does it make a difference which digit Isla uses? How about the digits that Zac and Aki use?*

STRENGTHEN Ask children to draw number lines for each interval of 1,000, to help them to visualise the position of each number.

DEEPEN Challenge children to explain and justify how to use the place value of the 100s digit to help when deciding whether to round up or down.

ASSESSMENT CHECKPOINT Question ❸ assesses whether children can round a given number to the nearest 1,000 by identifying which is the nearest multiple of 1,000.

ANSWERS Answers for the **Practice** part of the lesson can be found in the *Power Maths* online subscription.

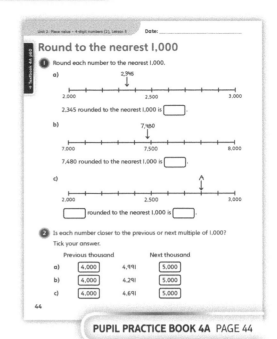

PUPIL PRACTICE BOOK 4A PAGE 44

PUPIL PRACTICE BOOK 4A PAGE 45

Reflect

WAYS OF WORKING Independent thinking

IN FOCUS The **Reflect** part of the lesson prompts children to explain the steps in their decision-making process.

ASSESSMENT CHECKPOINT Assess whether children can explain the process they go through when rounding a number to the nearest multiple of 1,000.

ANSWERS Answers for the **Reflect** part of the lesson can be found in the *Power Maths* online subscription.

After the lesson ⏸

- Can children use what they have learned to play a 'Guess my number' game?
  Say, for example: *I am thinking of a number that rounds to 2,000 to the nearest 1,000.*
  Children ask yes / no questions to guess the number.

PUPIL PRACTICE BOOK 4A PAGE 46

Round to the nearest 100

Learning focus

In this lesson, children round 3- and 4-digit numbers to the nearest 100.

Before you teach ⏸

- What contexts can children relate to so that their understanding of this concept is deepened?
- Based on previous lessons taught in this unit, are there any additional misconceptions you need to consider?

NATIONAL CURRICULUM LINKS

Year 4 Number – number and place value

Round any number to the nearest 10, 100 or 1,000.

ASSESSING MASTERY

Children can identify the 100 either side of a given number. They can successfully round to the nearest 100 and understand how the number line can help them to do this.

COMMON MISCONCEPTIONS

Children may struggle with rounding numbers to the nearest 100 when the answers are 0 and 1,000 (for example with numbers such as 34 or 984). Ask:
- *What are the 100s either side of 34? Try drawing this on a number line.*
- *What happens to the place value of the digits in 984 when you round to the nearest 100?*

STRENGTHENING UNDERSTANDING

To strengthen understanding, begin by asking children to focus on numbers that are in the range 0 to 1,000.

GOING DEEPER

Ask children to give examples of when they would need to round to the nearest 100. Ask: *What number should 550 be rounded to? Explain why.*

KEY LANGUAGE

In lesson: rounding, hundreds (100s)

Other language to be used by the teacher: round up, round down, nearest, thousands (1,000s), tens (10s), number line

STRUCTURES AND REPRESENTATIONS

Number lines

RESOURCES

Mandatory: blank number lines

Optional: dice to generate numbers, specifically numbered number lines

 In the eTextbook of this lesson, you will find interactive links to a selection of teaching tools.

Quick recap 🔄

Ask children to say three numbers that round to 1,000 to the nearest 1,000. Then ask them to say three numbers that round to 2,000. Then ask them to say three numbers that round to 9,000.

Discover

WAYS OF WORKING Pair work

ASK

- Question ❶ a): *Say the number out loud. Can you hear the number of 100s?*
- Question ❶ b): *If you partitioned the number into 100s, 10s and 1s, how would you say it?*

IN FOCUS Children are identifying the previous and the next multiple of 100 for given 3-digit numbers. They are then using this to round to the nearest multiple of 100.

PRACTICAL TIPS Provide a number line from 0 to 1,000 with intervals marked for each 100. As a class, count up together along the line in 100s.

ANSWERS

Question ❶ a): 568 is between 500 and 600.

Question ❶ b): 568 rounds to 600.

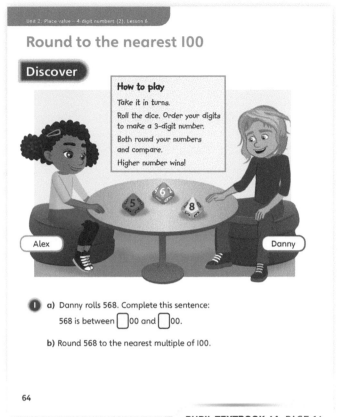

Share

WAYS OF WORKING Whole class teacher led

ASK

- Question ❶ a): *How can you prove that 568 is greater than 500?*
- Question ❶ a): *How can you prove that 568 is less than 600?*
- Question ❶ b): *Show why 550 is the half-way point between 500 and 600.*
- Question ❶ b): *Could you play this game yourself?*

IN FOCUS Question ❶ a) addresses rounding without stating that we are asking children to round. The use of number lines to see which 100 the number is closer to is reinforced from the previous lesson.

In question ❶ b), children need to recognise the nearest multiple of 100 in order to round the given number.

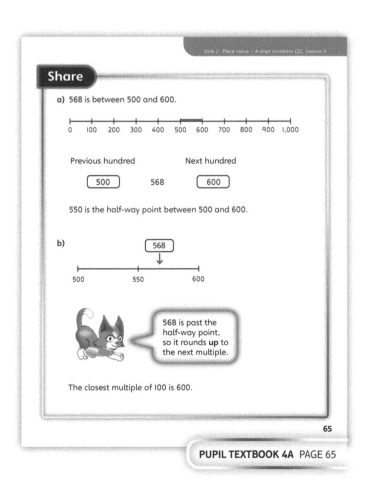

99

Think together

WAYS OF WORKING Whole class teacher led (I do, We do, You do)

ASK

- Question **1**: *What is the same and what is different about each number?*
- Question **2**: *What number line could you draw for part (a)? What about part (b), or part (c)?*
- Question **3**: *Do you notice any patterns? Which digits do you look at when rounding to the nearest 100?*

IN FOCUS Questions **1** and **2** provide children with a structure that builds on from the **Share** section. It is an opportunity for children to discuss how they know which hundred to round to. In question **2** children revisit what happens when they are presented with a number that has 5 tens.

STRENGTHEN Focus on numbers that are in the range 0 to 1,000 first, before then moving on to numbers in the range 1,000 to 2,000.

DEEPEN Ask children to explore the patterns in question **3** more fully. Can they create lists that are similar to A, B and C and then round them to the nearest 100?

ASSESSMENT CHECKPOINT In question **1**, assess whether children understand how to work out the 100s either side of the given numbers. In question **2** check that children are looking at the 10s digit in a number when rounding to the nearest 100.

ANSWERS

Question **1**: 812 → 800; 880 → 900; 857 → 900; 808 → 800; 850 → 900

Question **2** a): Ends of number line are 4,500 and 4,600; 4,595 → 4,600.

Question **2** b): Ends of number line are 2,300 and 2,400; 2,340 → 2,300.

Question **2** c): Ends of number line are 1,000 and 1,100; 1,050 → 1,100.

Question **3**:

List A	List B	List C
503 → 500	2,402 → 2,400	547 → 500
513 → 500	2,412 → 2,400	1,547 → 1,500
523 → 500	2,422 → 2,400	2,547 → 2,500
533 → 500	2,432 → 2,400	3,547 → 3,500
543 → 500	2,442 → 2,400	4,547 → 4,500
553 → 600	2,452 → 2,500	5,547 → 5,500
563 → 600	2,462 → 2,500	6,547 → 6,500
573 → 600	2,472 → 2,500	7,547 → 7,500
583 → 600	2,482 → 2,500	8,547 → 8,500
593 → 600	2,492 → 2,500	9,547 → 9,500

Lists A and B depend on the tens digits: 1, 2, 3 and 4 tens make the number round down and a tens digit 5 or above makes the number round up.

List C has the same hundreds, tens and ones (547), so they all round down to ☐,500 (where the thousands digit matches the original number).

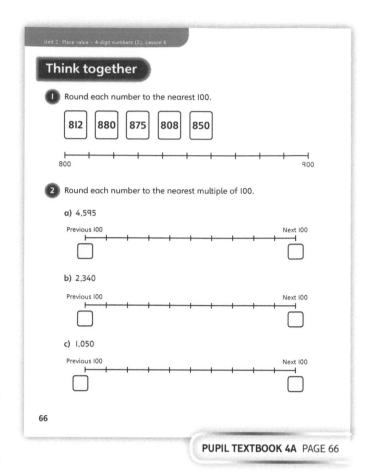

Think together

1 Round each number to the nearest 100.

812   880   875   808   850

800   900

2 Round each number to the nearest multiple of 100.

a) 4,595

Previous 100   Next 100

b) 2,340

Previous 100   Next 100

c) 1,050

Previous 100   Next 100

66

PUPIL TEXTBOOK 4A PAGE 66

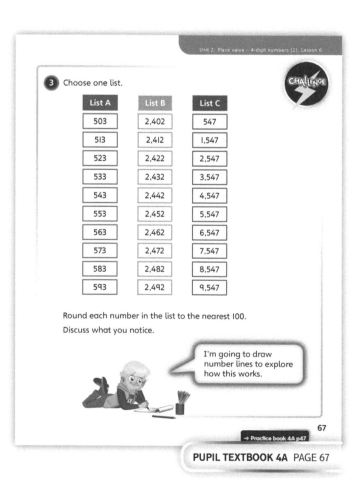

3 Choose one list.

CHALLENGE

List A	List B	List C
503	2,402	547
513	2,412	1,547
523	2,422	2,547
533	2,432	3,547
543	2,442	4,547
553	2,452	5,547
563	2,462	6,547
573	2,472	7,547
583	2,482	8,547
593	2,492	9,547

Round each number in the list to the nearest 100.

Discuss what you notice.

I'm going to draw number lines to explore how this works.

→ Practice book 4A p47

67

PUPIL TEXTBOOK 4A PAGE 67

Practice

WAYS OF WORKING Independent thinking

IN FOCUS Question ❶ reinforces the use of number lines to help with identifying which 100 a given number is closest to. Questions ❷ and ❸ become more abstract, requiring children to look at the 10s digit to help them round to the nearest 100.

Question ❹ requires children to explain what happens when there are 5 tens in the number. Questions ❺ and ❻ are problem-solving questions that require children to use what they know about place value and rounding to find mystery numbers.

STRENGTHEN Ask children to justify how they know which digit to round up or down and why. Is there a real-life context where they would need to round to the nearest 100?

DEEPEN Explore rounding to the nearest 100 when 0 or 1,000 is the answer. Ask children: *What does 987 round to? Which digits change? What is 34 rounded to the nearest 100? What is different about your answer?*

THINK DIFFERENTLY In question ❹ children are asked to think about two possible answers to a problem and then to say whether they think either of the answers are correct and to explain their reasoning.

ASSESSMENT CHECKPOINT Question ❶ allows you to assess whether children can use a number line to round numbers by finding the nearest 100. The rest of the questions should allow you to see if children can work in a more abstract way to round a number to the nearest 100. Check that children are able to identify the 100s either side of the given number and that they are looking at the 10s digit to round up or down.

ANSWERS Answers for the **Practice** part of the lesson can be found in the *Power Maths* online subscription.

Reflect

WAYS OF WORKING Independent thinking

IN FOCUS The **Reflect** part of the lesson prompts children to discuss whether to round numbers up or down. This will prepare them to explore this concept in more detail in subsequent lessons.

ASSESSMENT CHECKPOINT Assess whether children can justify their reasoning for rounding up or rounding down.

ANSWERS Answers for the **Reflect** part of the lesson can be found in the *Power Maths* online subscription.

After the lesson ⏸

• Were children confident using a number line?
• Can children use a number line to round numbers to the nearest 100?
• Are children able to round numbers to the nearest 100 without a number line?

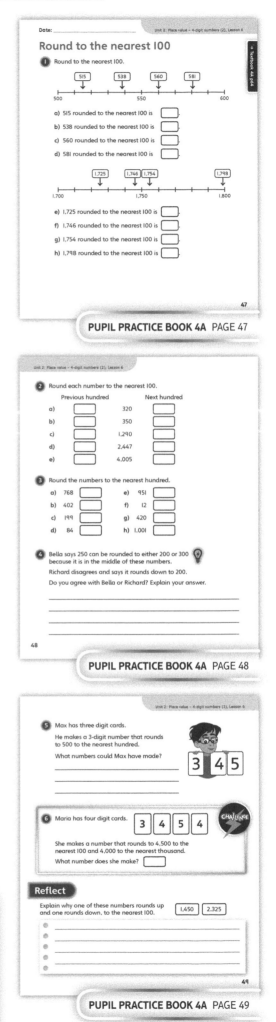

PUPIL PRACTICE BOOK 4A PAGE 47

PUPIL PRACTICE BOOK 4A PAGE 48

PUPIL PRACTICE BOOK 4A PAGE 49

Round to the nearest 10

Learning focus

In this lesson, children are rounding to the nearest multiple of 10.

Before you teach ⏸

- Do children know what multiples of 10 are?
- Can children count up and down in 1s up to 1,000?
- Can children describe a number as 100s, 10s and 1s?

NATIONAL CURRICULUM LINKS

Year 4 Number – number and place value

Round any number to the nearest 10, 100 or 1,000.

ASSESSING MASTERY

Children can round any 2-, 3- or 4-digit number to the nearest 10 and understand the multiples of 10 above and below a specific number. Children understand when rounding to the nearest 10 that they need to look at the number of 1s to decide whether to round up or down.

COMMON MISCONCEPTIONS

Children may not be able to work out which multiples of 10 are above and below a number. Ask:
- *What is the same about multiples of 10? (They all end in a 0.)*
- *Can you count up and down to find the closest multiples of 10.*

Children may have difficulty deciding whether to round a number up or down. Ask:
- *Which multiple of 10 is your number closest to?*
- *If your number ends in 5 ones, do we round it up or down?*

STRENGTHENING UNDERSTANDING

To strengthen understanding, start by focussing on numbers in the range 0 to 100 in order to build children's confidence before moving on to bigger numbers.

GOING DEEPER

Ask children to start investigating the rules for rounding based on the value of the 1s digit. What patterns do they notice?

KEY LANGUAGE

In lesson: nearest, rounding, round up, round down, hundreds (100s), tens (10s), ones (1s), round up, round down

Other language to be used by the teacher: multiple of 10

STRUCTURES AND REPRESENTATIONS

Number lines

RESOURCES

Mandatory: number lines

Optional: digit cards, sticky notes

 In the eTextbook of this lesson, you will find interactive links to a selection of teaching tools.

Quick recap ↻

Ask children to say three numbers that round to 100 to the nearest 100. Then ask them to say three numbers that round to 1,100. Then ask them to say three numbers that round to 1,900.

Discover

Round to the nearest 10

WAYS OF WORKING Pair work

ASK

- Question ① a): *What is the same and what is different about the numbers on the cards?*
- Question ① b): *Can you partition each number into 10s and 1s?*

IN FOCUS Children are rounding a 2-digit number to the nearest 10, by identifying which is the closest multiple of 10.

PRACTICAL TIPS Use numbers on cards or sticky notes that can be positioned on a number line and then moved and arranged in order as required.

ANSWERS

Question ① a): Closer to 20: 21, 22, 23, 24.
Closer to 30: 26, 27, 28, 29.
25 is midway between 20 and 30.

Question ① b): 21, 22, 23 and 24 round down to 20.
25, 26, 27, 28 and 29 round up to 30.

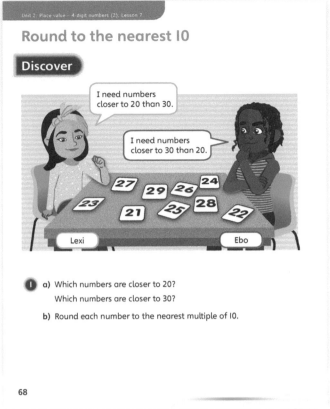

Round to the nearest 10

Discover

① a) Which numbers are closer to 20?
Which numbers are closer to 30?

b) Round each number to the nearest multiple of 10.

68

PUPIL TEXTBOOK 4A PAGE 68

Share

WAYS OF WORKING Whole class teacher led

ASK

- Question ① a): *Why is 25 an important number on this number line?*
- Question ① b): *Can you prove that 25 is exactly half-way between 20 and 30?*

IN FOCUS In question ① a), children are identifying the previous and the next multiple of 10 for given numbers and then using this to find which is the closest multiple of 10. This supports the concept of rounding.

Question ① b) requires children to then use this knowledge to round each number up or down to the nearest multiple of 10.

STRENGTHEN Discuss how the number line helps us to see which 10 the number is closer to, and encourage children to prove this by working out the subtractions to check their answers. First discuss with children how they can find the 10s that the number lies between. Point out the number of 10s in the number and help them to work out the 10s above and below this.

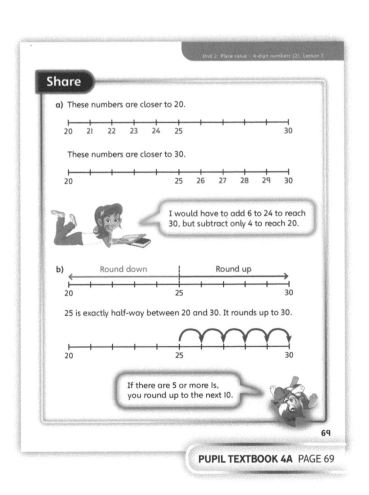

Share

a) These numbers are closer to 20.

These numbers are closer to 30.

I would have to add 6 to 24 to reach 30, but subtract only 4 to reach 20.

b) Round down | Round up

25 is exactly half-way between 20 and 30. It rounds up to 30.

If there are 5 or more 1s, you round up to the next 10.

69

PUPIL TEXTBOOK 4A PAGE 69

Think together

WAYS OF WORKING Whole class teacher led (I do, We do, You do)

ASK

- Question ❶: *What number line could you draw for each number?*
- Question ❷: *Does looking at the digits help with rounding?*
- Question ❸: *Can you think of a rule for whether to round up or down?*

IN FOCUS   In questions ❶, and ❷, children round 2-, 3- and 4-digit numbers to the nearest 10. When playing the game in question ❸, children will need to decide whether to round up or down.

STRENGTHEN   Continue to use the number line throughout to strengthen children's understanding of the multiples of 10 either side of a number and how close the number is to a multiple of 10. Also, encourage children to carry out simple subtractions to check their answers. Some children may look at patterns, such as: 21 – 20 = 1; 22 – 20 = 2; 23 – 20 = 3, alongside 30 – 21 = 9; 30 – 22 = 8; 30 – 23 = 7.

DEEPEN   Ask children to use what they have learned about rounding to play a 'Guess my number' game together. One child says: *I am thinking of a number that rounds to x to the nearest 10.* Others identify what the number could be.

ASSESSMENT CHECKPOINT   Use question ❷ to assess whether children can identify the nearest multiple of 10 for 2- and 3-digit numbers.

ANSWERS

Question ❶ a): Ends of the number line are 30 and 40; 32 → 30.

Question ❶ b): Ends of the number line are 160 and 170; 167 → 170.

Question ❶ c): Ends of the number line are 1,270 and 1,280; 1,278 → 1,280.

Question ❷: 48 rounds up to **50**.
131 rounds down to **130**.
40 stays at **40**.
55, **56**, **57**, **58**, **59**, **60**, **61**, **62**, **63** or **64** round to 60.

Question ❸: Answers will vary.
There are 12 two-digit numbers that you can make.
There are 24 three-digit numbers that you can make.
There are 24 four-digit numbers that you can make.
In each case we are rounding to the nearest 10, so if the digit in the 1s is 4 or less, the number rounds down to the previous 10. If the digit is 5 or more, the number rounds up to the next 10.

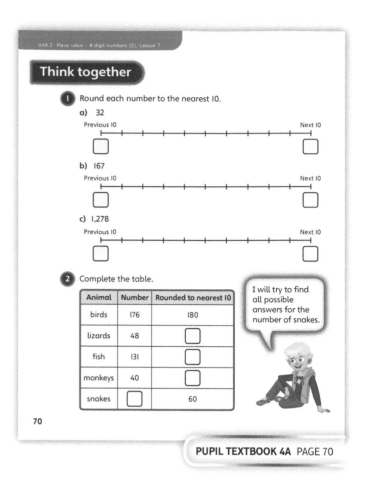

PUPIL TEXTBOOK 4A PAGE 70

PUPIL TEXTBOOK 4A PAGE 71

Practice

WAYS OF WORKING Independent thinking

IN FOCUS In question **1**, children are rounding to the nearest 10 using the number line to support their thinking. Questions **2** and **3** require children to round 2- and 3-digit numbers to the nearest 10. The number line model is no longer provided, but children may still find it helpful to draw their own.

In question **4**, children are focusing on the 1s digits and the idea of which digits round up or down.

In question **5**, children are rounding with 2-, 3- or 4-digit numbers and in questions **6** and **7** they must select digit cards to make numbers, and then choose whether to round up or down.

STRENGTHEN Give children blank number lines with intervals, to encourage them to continue to position the numbers. Encourage children to do the simple subtractions that will reinforce which 10 the number is closer to.

DEEPEN Ask children to find 4-digit numbers in scientific or historical fact files. Can they explain how to round each of their numbers to the nearest 10?

ASSESSMENT CHECKPOINT Question **5** assesses whether children can use their knowledge of place value and of multiples of 10 to round 2-, 3- and 4-digit numbers to the nearest 10.

ANSWERS Answers for the **Practice** part of the lesson can be found in the *Power Maths* online subscription.

Reflect

WAYS OF WORKING Independent thinking

IN FOCUS The **Reflect** part of the lesson prompts children to consider what they know about whether to round a number up or to round down.

ASSESSMENT CHECKPOINT Assess whether children can generate 4-digit numbers that round up or round down as required.

ANSWERS Answers for the **Reflect** part of the lesson can be found in the *Power Maths* online subscription.

After the lesson

- Did children use the number line in order to find the nearest multiples of 10 and then round to the nearest 10?
- Were children able to understand the importance of the 1s digit when rounding to the nearest 10 and understand that the 1s digit determines whether to round up or down?

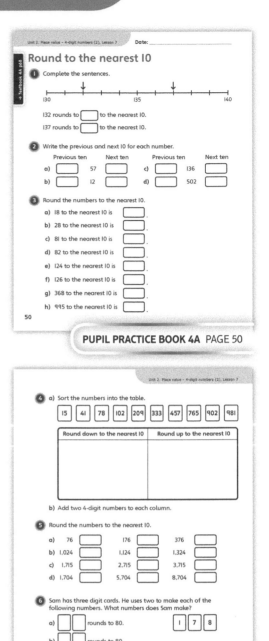

PUPIL PRACTICE BOOK 4A PAGE 50

PUPIL PRACTICE BOOK 4A PAGE 51

PUPIL PRACTICE BOOK 4A PAGE 52

Round to the nearest 1,000, 100 or 10

Learning focus

In this lesson, children will build on their knowledge of rounding to 1,000, 100 and 10, including working out numbers that round to a particular degree of accuracy.

Before you teach ⏸

- Can children round to the nearest 10?
- Can children round to the nearest 100?
- Can children round to the nearest 1,000?

NATIONAL CURRICULUM LINKS

Year 4 Number – number and place value

Round any number to the nearest 10, 100 or 1,000.

ASSESSING MASTERY

Children can successfully use their previous learning of rounding to the nearest 10, 100 and 1,000 to answer a variety of problems, including problems where they must generate numbers that round to a particular degree of accuracy.

COMMON MISCONCEPTIONS

Children may not realise that a number can round either up or down, depending on the degree of accuracy. For example 7,189 could round up to 7,190 (to the nearest 10) or to 7,200 (to the nearest 100), but also down to 7,000 (to the nearest 1,000). Show a suitable number line and discuss the position of the number on it (for example, the position of 7,189 on a number line from 7,000 to 7,200).

STRENGTHENING UNDERSTANDING

Children may need further practice rounding numbers that are (or are within a few increments of) half-way between two numbers. Ensure that children have access to concrete equipment and a place value grid and ask questions such as:
- *What is the difference between 6,499 and 6,501 when rounding? How do you know? What place value digits is it important to consider here? What does 6,500 round to, to the nearest 1,000?*

Model the answers as you go, if necessary, to consolidate understanding.

GOING DEEPER

To prompt deeper thinking in this lesson, ask a series of structured questions such as:
- *A number rounded to the nearest 1,000/100/10 is 3,000. What could the number be? How many numbers can you find?* What is the greatest number? What is the smallest number?

To extend thinking further, ask:
- *A number rounded to the nearest 1,000 is 3,000. The same number rounded to the nearest 100 is 3,100. What could the* number *be? How many numbers can you find?*

Children who are confident and secure answering these questions could make up their own questions to challenge each other.

KEY LANGUAGE

In lesson: rounds to, nearest, number line, closer to, greatest, smallest, hundreds (100s)

Other language to be used by the teacher: round up, round down

STRUCTURES AND REPRESENTATIONS

Number lines, place value grids

RESOURCES

Mandatory: number lines, place value grid

Optional: base 10 equipment, counters

 In the eTextbook of this lesson, you will find interactive links to a selection of teaching tools.

Quick recap 🔄

Ask children to say three numbers that round to 10 to the nearest 10. Then ask them to say three numbers that round to 110. Then ask them to say three numbers that round to 190.

Discover

Unit 2: Place value – 4-digit numbers (2), Lesson 8

WAYS OF WORKING Pair work

ASK

- Question **1** a): *How many grams are in a kilogram? How might this help you?*
- Question **1** a): *What number is half-way between 4,000 and 5,000? Will this number be closer to 4,000 or 5,000? How do you know?*

IN FOCUS Question **1** a) requires children to estimate the position of a number on the number line. They should consider the value of each digit when placing the number. Can they explain why the number is so much closer to 5,000 than to 4,000?

Question **1** b) focuses on helping children understand what it is important to look at when rounding to the nearest 1,000, 100 or 10. Children should first consider the 4 thousands. Some children may think that this means the number cannot round to 5,000. Encourage children to look at the 100s. Do they realise that because there are 9 hundreds, the number rounds up to 5,000?

PRACTICAL TIPS Allow children to use concrete equipment such as number lines, place value counters and a place value grid so that they can physically represent the question on their classroom tables.

ANSWERS

Question **1** a):

4,000 g   4,949   5,000 g

Question **1** b): Nearest 1,000: 5,000; nearest 100: 4,900; nearest 10: 4,950

Share

WAYS OF WORKING Whole class teacher led

ASK

- Question **1** b): *Do you think it will round to the same number each time? Why? Why not?*

IN FOCUS For question **1** b), have children share their thoughts and their methods. Discuss the strategy of using the number line to identify the previous and next 1,000, 100 or 10 and finding which the number is closer to. Ensure children notice that the answer is different each time. Look for clear explanations and share these with the class.

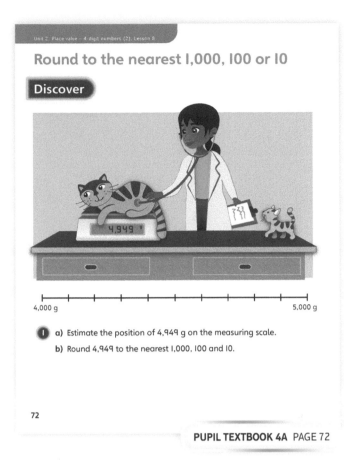

Round to the nearest 1,000, 100 or 10

Discover

1 a) Estimate the position of 4,949 g on the measuring scale.

    b) Round 4,949 to the nearest 1,000, 100 and 10.

72

PUPIL TEXTBOOK 4A PAGE 72

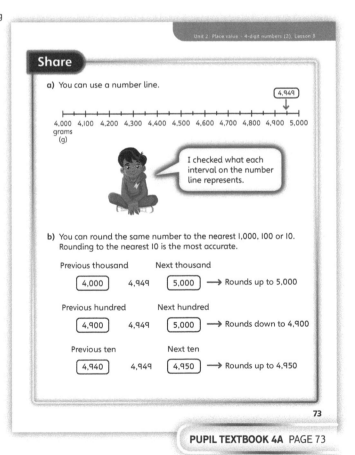

Share

a) You can use a number line.

I checked what each interval on the number line represents.

b) You can round the same number to the nearest 1,000, 100 or 10. Rounding to the nearest 10 is the most accurate.

Previous thousand	Next thousand	
4,000 4,949	5,000	→ Rounds up to 5,000

Previous hundred	Next hundred	
4,900 4,949	5,000	→ Rounds down to 4,900

Previous ten	Next ten	
4,940 4,949	4,950	→ Rounds up to 4,950

73

PUPIL TEXTBOOK 4A PAGE 73

Think together

WAYS OF WORKING Whole class teacher led (I do, We do, You do)

ASK

- Question **1**: *What number is shown? What is the value of each digit? How could you round without using a number line?*
- Question **2**: *What can you say about the number of 1,000s in each of the numbers? Is there more than one answer? Why?*
- Question **3**: *What information do you know? Can you use counters to help you? Can you explain your methods? Which do you think is best? Why?*

IN FOCUS Question **1** is similar to the **Discover** task, practising the skills of rounding a number to the nearest 10, 100 and 1,000.

In question **2**, children identify the range of numbers that will round to a given multiple of 1,000.

Question **3** requires children to discuss several possible methods for rounding a given number, including using a number line, identifying the previous and next multiples, and considering the 100s digit in particular. Look for children referring to the value of each digit in the number and explaining how each method helps them see whether to round up or round down, and what number to round to.

STRENGTHEN Allow children to use a number line as they work through this section. In question **3**, children can also use counters to make the number.

DEEPEN To extend thinking in this lesson, ask children to make numbers that round to different degrees of accuracy. For example, can they write three numbers that round to 3,400 to the nearest 100? Then give them two conditions, and then three conditions that they must work to. Can children show the range of applicable values?

ASSESSMENT CHECKPOINT Assess whether children can round a number to the nearest 10, 100 and 1,000 by considering the size of the digits. Can children work out numbers that round to a particular degree of accuracy?

ANSWERS

Question **1**: Nearest 10: 3,510; nearest 100: 3,500; nearest 1,000: 4,000

Question **2**: **1,500** is the smallest number to round (up) to 2,000.

2,499 is the greatest number to round (down) to 2,000.

Question **3**: Drawing a number line can be helpful but is time consuming and a different line is needed for each rounding, depending on what the number is being rounded to.

Working out the previous and next multiples is a useful method as long as children can work out which multiples are appropriate.

Looking at the digits is a good method for all rounding as long as children know which digit to look at.

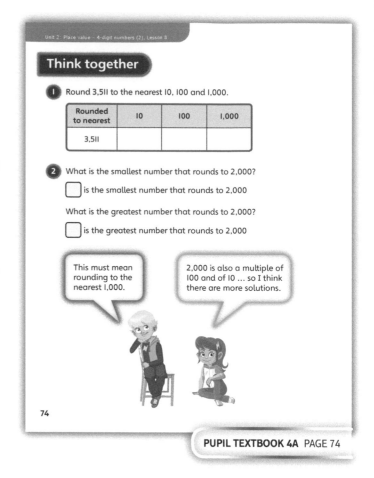

PUPIL TEXTBOOK 4A PAGE 74

PUPIL TEXTBOOK 4A PAGE 75

Practice

WAYS OF WORKING Independent thinking

IN FOCUS Questions in this **Practice** section focus on practising rounding to the nearest 10, 100 and 1,000. Children start by rounding to the nearest 1,000 in question ❶, before rounding to the nearest 100 in question ❷ and the nearest 10 in question ❸. In question ❹, children explore the fact that a number can sometimes round up or down to different numbers, depending on the degree of accuracy required in the question. In question ❺ they identify the upper and lower limits of numbers that will round to a given degree of accuracy. In question ❻ children apply reasoning to find different solutions to a rounding problem, and in question ❼ they work through a missing digits problem.

STRENGTHEN Place value grids and concrete equipment such as base 10 equipment or place value counters can be used to visually represent the numbers in the word problems.

DEEPEN Explore thinking more deeply in question ❼ by asking children how they found the missing digits. What do they know about each of the missing digits? What information does each column of the table give them? Could there ever be more than one possible answer?

THINK DIFFERENTLY Question ❸ encourages a different way of thinking about rounding. Children can use their knowledge of place value to round numbers with 2-, 3- and 4-digits to the nearest 10..

In question ❻, children use their knowledge of rounding to the nearest 10, 100 and 1,000 to identify a number that satisfies three conditions. Children may reason that they could simply find a number that rounds to 2,000 to the nearest 10, as it would automatically round to 2,000 to the nearest 100 and to the nearest 1,000.

ASSESSMENT CHECKPOINT Assess whether children can solve a range of problems in a variety of contexts that involve rounding. Can they demonstrate that they can round numbers to the nearest 10, 100 and 1,000 and are they able to write numbers that round to a particular degree of accuracy?

ANSWERS Answers for the **Practice** part of the lesson can be found in the *Power Maths* online subscription.

Reflect

WAYS OF WORKING Pair work

IN FOCUS This question brings together all aspects of rounding from the lesson. Children refer to a representation of a 4-digit number on a place value grid and explain how it rounds to the nearest 10, 100 and 1,000. Look out for children considering the place value of each digit in the number in order to round accurately.

ASSESSMENT CHECKPOINT Assess whether children can explain how to round to a specified degree of accuracy.

ANSWERS Answers for the **Reflect** part of the lesson can be found in the *Power Maths* online subscription.

After the lesson ⏸

- Can children round to the nearest 10, 100 and 1,000 using a number line?
- Can children identify numbers that round to a particular number and degree of accuracy?
- Can children solve rounding problems in a range of contexts?

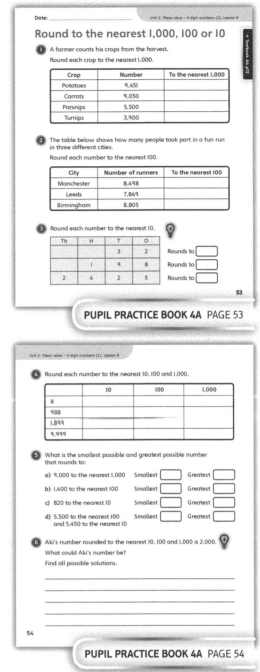

PUPIL PRACTICE BOOK 4A PAGE 53

PUPIL PRACTICE BOOK 4A PAGE 54

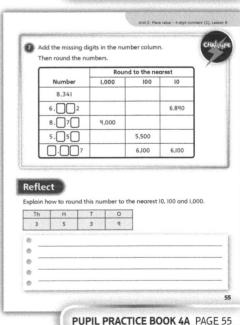

PUPIL PRACTICE BOOK 4A PAGE 55

End of unit check

Don't forget the unit assessment grid in your *Power Maths* online subscription.

WAYS OF WORKING Group work adult led

IN FOCUS Children recap their understanding of ordering and rounding 4-digit numbers and representing numbers on number lines. Question ① focuses on recognising a number on a number line. Question ② looks at rounding to the nearest 1,000. Question ③ looks at rounding a 4-digit number to the nearest 100. Question ④ focuses on comparing numbers which are shown in different pictorial representations. Question ⑤ focuses on ordering four masses. Question ⑥ is a SATS-style question focusing on rounding to the nearest 10, 100 and 1,000.

ANSWERS AND COMMENTARY Children who have mastered the concepts in this unit will know that a 4-digit number is made up of 1,000s, 100s, 10s and 1s and will be able to compare and order 4-digit numbers by looking at the digits in each place value column. Children will understand the number line to 10,000 and will be beginning to know where numbers lie on the number line. Children will be able to round numbers to the nearest 10, 100 and 1,000.

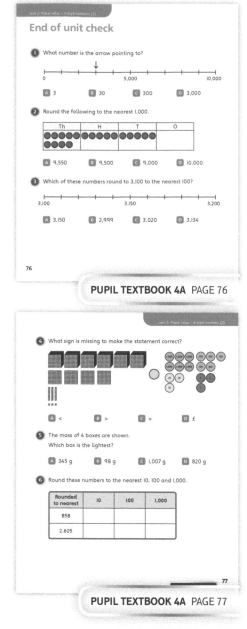

PUPIL TEXTBOOK 4A PAGE 76

PUPIL TEXTBOOK 4A PAGE 77

Q	A	WRONG ANSWERS AND MISCONCEPTIONS	STRENGTHENING UNDERSTANDING
1	D	Choosing A, B or C indicates that the child has not understood the scale on the number line.	Secure children's understanding of 4-digit numbers by using base 10 equipment and place value counters in conjunction with a place value grid.
2	D	Choosing A indicates that the child has found the number represented. B suggests that the child has rounded down incorrectly to the nearest 100. C suggests that the child has rounded down instead of up.	When comparing 3-digit and 4-digit numbers, use the place value grid to show the importance of comparing the same place value, not the first digit of the number.
3	D	Choosing A suggests the child incorrectly rounds down. Choosing B or C indicates greater strengthening of how to round is needed.	
4	A	Choosing B or C suggests that the child has not interpreted each pictorial representation correctly.	Link physical numbers to numbers on the number line to support children with counting.
5	B	Choosing any other answer suggests that the child needs strengthening of their understanding of place value.	Encourage children to count aloud when counting on or back on a number line to help them see the counting pattern.
6	858: 860; 900; 1,000 2,605: 2,610; 2,600; 3,000	2,605 rounds up to 2,610 (nearest 10) because the 1s digit is 5, but rounds down to 2,600 (nearest 100) because the 10s digit is 0.	

My journal

WAYS OF WORKING Independent thinking

ANSWERS AND COMMENTARY

In question **1**, children should recall that you consider the digit in the 100s column when rounding to the nearest 1,000. If the digit is less than 5, the number rounds down to the previous 1,000. If the number is 5 or greater, the number rounds up to the next 1,000.

In question **2**, children will find that when they round a number to the nearest 1,000 all the columns can change; for example, 7,564 to the nearest 1,000 is 8,000.

Power check

WAYS OF WORKING Independent thinking

ASK

• *How confident do you feel about ordering and comparing 4-digit numbers?*
• *Do you think you could explain how to find 1,000 more or less than a number?*
• *Would you feel confident rounding to the nearest 10, 100 or 1,000?*

Power play

WAYS OF WORKING Pair work or small groups

IN FOCUS Use this **Power play** to check if children understand the different objectives covered in the unit. Children focus on rounding to the nearest 10, 100 and 1,000 and finding 100 more and 1,000 more or less than a number. The modification suggested by Sparks requires children to think carefully about where they place the digits and the importance of the order.

ANSWERS AND COMMENTARY Within this **Power play**, children will generate a variety of numbers and show whether they can round them correctly and find 100 more and 1,000 more or less.

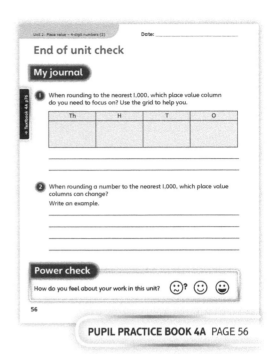

PUPIL PRACTICE BOOK 4A PAGE 56

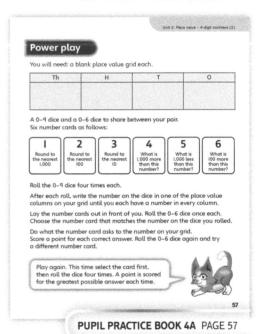

PUPIL PRACTICE BOOK 4A PAGE 57

After the unit ⏸

• Can children represent 4-digit numbers on a number line?
• Are children able to show a good knowledge of place value in order to compare and order numbers?

Strengthen and **Deepen** activities for this unit can be found in the *Power Maths* online subscription.

Unit 3
Addition and subtraction

Mastery Expert tip! 'The bar model is useful for representing additions and subtractions, but don't just show it to children, ask them to explain it and to draw their own!'

Don't forget to watch the Unit 3 video!

WHY THIS UNIT IS IMPORTANT

This unit is important because it focuses on learning a range of addition and subtraction strategies, in particular the column method. Mastering this will lead to confidence in many other areas of mathematics, especially when children apply their strategies to word problems and puzzles.

WHERE THIS UNIT FITS

→ Unit 2: Place value – 4-digit numbers (2)

→ **Unit 3: Addition and subtraction**

→ Unit 4: Measure – area

This unit builds on children's Year 3 work on adding and subtracting with 3-digit numbers. It further develops their estimation and answer-checking strategies and their problem-solving skills. This unit provides essential preparation for beginning to add and subtract numbers with more than four digits.

Before they start this unit, it is expected that children:
- have a firm understanding of place value (up to 4-digit numbers)
- know a range of mental addition and subtraction strategies
- can apply these strategies to a range of contexts including measure.

ASSESSING MASTERY

Children who have mastered this unit can find totals and differences using the column method of addition and subtraction. They should not, however, always rely on the column method, but should understand when there is a more efficient method. They can confidently apply their knowledge when solving word problems and explain all answers clearly, using the correct vocabulary.

COMMON MISCONCEPTIONS	STRENGTHENING UNDERSTANDING	GOING DEEPER
Children may not align the columns correctly when using the column method.	Run an intervention in which children use place value grids to support aligning columns and understanding the importance of this.	Solve some addition and subtraction sentences that have missing numbers. Provide children with some multi-step word problems. Can they represent them with a diagram and then solve them?
Children may not understand the place value behind the method of exchanging.	Practise exchanging using place value grids and place value counters.	Ask children to make up their own word problems to fit an addition or subtraction sentence.
Children may not know whether to add or subtract when solving a problem.	Ask children to represent the problem with a bar model.	

Unit 3: Addition and subtraction

Go through the unit starter pages of the **Pupil Textbook**. Talk through the key learning points that the characters mention and the key vocabulary.

STRUCTURES AND REPRESENTATIONS

Place value grid: This model uses counters to show the value of each column, which supports the column method layout.

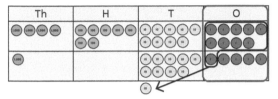

Bar model: This model can be used to represent the situation in some addition and subtraction word problems.

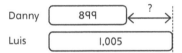

Danny [899] ← ? →

Luis [1,005]

Part-whole model: This model is an alternative way to represent the situation in addition and subtraction word problems.

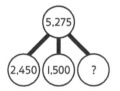

5,275

2,450   1,500   ?

KEY LANGUAGE

There is some key language that children will need to know as a part of the learning in this unit.

→ addition, subtraction
→ total
→ more than, less than
→ difference, exchange
→ column method
→ estimate, accurate, efficient, exact
→ strategy
→ diagram
→ how much
→ fact

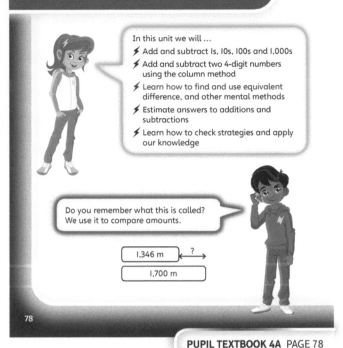

Unit 3
Addition and subtraction

In this unit we will …
⚡ Add and subtract 1s, 10s, 100s and 1,000s
⚡ Add and subtract two 4-digit numbers using the column method
⚡ Learn how to find and use equivalent difference, and other mental methods
⚡ Estimate answers to additions and subtractions
⚡ Learn how to check strategies and apply our knowledge

Do you remember what this is called? We use it to compare amounts.

1,346 m ← ? →
1,700 m

78

PUPIL TEXTBOOK 4A PAGE 78

We will need some maths words. Do you know what they all mean?

addition	total	more than
subtraction	less than	column method
estimate	how much	strategy
efficient	accurate	exact	fact

We need to use the part-whole model too. It helps us to break down and solve problems.

3,700 g

1,000 g   500 g   ?

79

PUPIL TEXTBOOK 4A PAGE 79

Add and subtract 1s, 10s, 100s, 1,000s

Learning focus

In this lesson, children will use their knowledge of place value to add and subtract 1, 10, 100 and 1,000 to and from 4-digit numbers.

Before you teach

- Would base 10 equipment help some children with their understanding of place value in this lesson?
- Which children do you think will need support in this lesson?

NATIONAL CURRICULUM LINKS

Year 4 Number – addition and subtraction

Add and subtract numbers with up to four digits using the formal written methods of columnar addition and subtraction where appropriate.

Year 4 Number – number and place value

Solve number and practical problems that involve addition and subtraction with increasingly large positive numbers.

ASSESSING MASTERY

Children can quickly make mental calculations when adding and subtracting 1s, 10s, 100s and 1,000s. Children can explain their method, demonstrating a deep understanding of place value, and can solve related problems in a range of contexts.

COMMON MISCONCEPTIONS

Children may have place value misconceptions, i.e. they may think 3,423 + 100 = 4,423. Ask:
- *Can you put the numbers into a place value grid to help?*

STRENGTHENING UNDERSTANDING

Give children the opportunity to practise adding 1, 10, 100, 1,000 to a range of numbers with a place value grid to help. Repeat until children can calculate place value additions and subtractions mentally.

GOING DEEPER

Give children a variety of missing number place value problems which will require them to think more deeply about the relationship between digits and what information they can use to find the missing numbers. For example, 3,487 + 2,000 = ☐, 1,298 − 70 = ☐, 6,815 + ☐ = 6,819, 2,731 − ☐ = 2,131.

KEY LANGUAGE

In lesson: more, fact, add, addition, subtract, subtraction

Other language to be used by the teacher: place value, thousands (1,000s), hundreds (100s), tens (10s), ones (1s), add, subtract, reduce, increase

STRUCTURES AND REPRESENTATIONS

Place value grid

RESOURCES

Mandatory: place value counters, base 10 equipment

 In the eTextbook of this lesson, you will find interactive links to a selection of teaching tools.

Quick recap

Play 'Say the next number'. Write or say a 3-digit number and ask children to call out or write the next number, focusing on using a counting strategy. Repeat with more 3-digit numbers. Play the game again, but this time ask children to say the previous number.

Discover

WAYS OF WORKING Pair work

ASK

- Question ① a): *What is the same about 3, 30, 300 and 3,000? What is different?*
- Question ① b): *Can you explain your answer?*

IN FOCUS Ask children what is similar and what is different about +3, +30, +300 and +3,000. Doing this will help them to think about the relationship and connections between these numbers.

PRACTICAL TIPS For this activity, some children may benefit from representing the numbers in the place value grids with concrete objects – use base 10 equipment for this.

ANSWERS

Question ① a):

Th	H	T	O
(1,000)(1,000)(1,000)(1,000)	(100)(100)	(10)(10)(10)(10)(10)(10)	(1)(1)(1)(1)(1)(1) (1)

Question ① b): 4,256 + 3 = 4,259
4,256 + 30 = 4,286
4,256 + 300 = 4,556
4,256 + 3,000 = 7,256

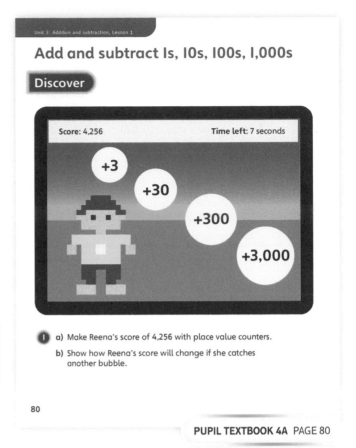

Add and subtract 1s, 10s, 100s, 1,000s

Discover

I a) Make Reena's score of 4,256 with place value counters.

b) Show how Reena's score will change if she catches another bubble.

PUPIL TEXTBOOK 4A PAGE 80

Share

WAYS OF WORKING Whole class teacher led

ASK

- Question ① a): *How many 1,000s are there? How many 100s? 10s? 1s?*
- Question ① b): *Can you explain what has happened in each of the answers? Which digits change? Which digits stay the same?*

IN FOCUS The place value grids support children with their understanding of place value. Count the counters in the grids aloud as a whole class – doing this will help children understand the numbers at a deeper level, and will also help them to make comparisons between numbers.

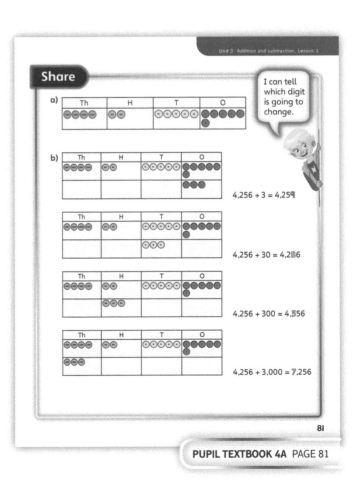

PUPIL TEXTBOOK 4A PAGE 81

Think together

WAYS OF WORKING Whole class teacher led (I do, We do, You do)

ASK

- Question **1**: *Which digit is changing in each calculation – the 1s, 10s, 100s or 1,000s?*
- Question **1**: *Why do you need 0s in numbers like 1,001?*
- Question **2**: *How can you work out calculations with missing numbers?*

IN FOCUS In question **1**, you may need to highlight that in the second example there are no 10s left, and so you need to include 0 as a placeholder, i.e. 7,646 – 40 = 7,606. Some children may not understand how to write this and so may give 766 as their answer.

STRENGTHEN For each question, provide base 10 equipment to visually represent the place value of the digits in the numbers for children who need it.

Asking children to explain their working will strengthen learning.

DEEPEN Give children some calculations with mistakes, for example 4,576 – 30 = 4,276. Ask: *Can you find the mistakes? Can you explain what the mistakes are?*

ASSESSMENT CHECKPOINT Use question **2** to assess whether children can work mentally, or whether they still rely on place value equipment.

ANSWERS

Question **1** a): 7,646 – 4 = **7,642**

Question **1** b): 7,646 – 40 = **7,606**

Question **1** c): 7,646 – 400 = **7,246**

Question **1** d): 7,646 – 4,000 = **3,646**

Question **2** a): 8,888 – 500 = **8,388**

Question **2** b): 8,888 – **5** = 8,883

Question **2** c): **3,888** = 8,888 – 5,000

Question **2** d): 8,838 = 8,888 – **50**

Question **3** a): 6,869 points

Question **3** b): There are many solutions for this question, as long as the star and the bubble lead to a score increase of 10. Examples include: a –10 star, then a +20 bubble, a –20 star, then a +30 bubble, and so on.

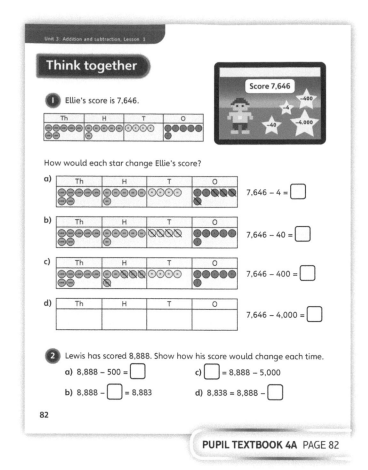

PUPIL TEXTBOOK 4A PAGE 82

PUPIL TEXTBOOK 4A PAGE 83

Practice

IN FOCUS Make sure children understand the contexts in question **4**, what the prices were originally and how they have now been reduced.

STRENGTHEN Question **6** will strengthen learning by encouraging children to think about place value relationships across a variety of numbers. The task may seem difficult at first, but reassure children that if they think hard they can reach a solution. Build children's confidence by explaining that there are multiple answers for each question.

DEEPEN Deepen learning by providing two-step questions with missing numbers, for example, 4,264 + ☐ − 200 = 4,564.

THINK DIFFERENTLY Question **5** challenges children to relate addition and subtraction. Listen carefully to children's reasoning for this question.

ASSESSMENT CHECKPOINT Question **4** will allow you to assess which children are able to apply their knowledge in context. Children should demonstrate problem-solving skills to work with what they know and complete the steps needed to find the solution.

ANSWERS Answers for the **Practice** part of the lesson can be found in the *Power Maths* online subscription.

Reflect

IN FOCUS This section will give children the opportunity to explain their understanding of the lesson. Encourage them to use a place value grid and place value counters as part of their answer.

ASSESSMENT CHECKPOINT Can children explain the method correctly? Do they use the correct vocabulary?

ANSWERS Answers for the **Reflect** part of the lesson can be found in the *Power Maths* online subscription.

After the lesson

- How will you support children who found the learning difficult in this lesson?
- What intervention sessions would be useful?
- Which children mastered the lesson?
- Could you make a display to support children in the subsequent lessons?

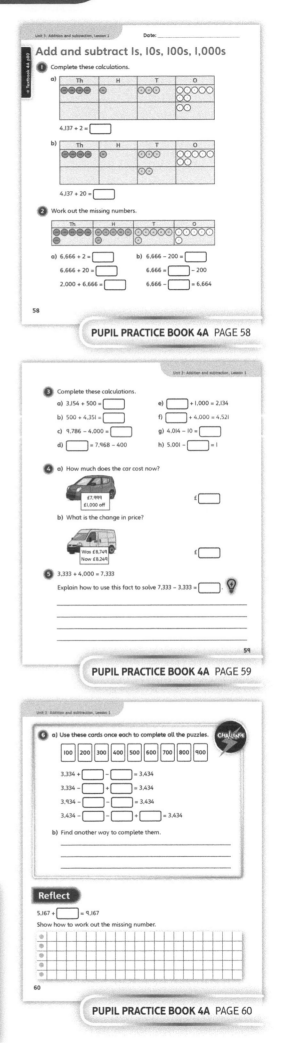

PUPIL PRACTICE BOOK 4A PAGE 58

PUPIL PRACTICE BOOK 4A PAGE 59

PUPIL PRACTICE BOOK 4A PAGE 60

Add two 4-digit numbers

Learning focus

In this lesson, children will add 4-digit numbers using the column method (without exchanging). This is closely paired with a place value grid to ensure children have a deeper understanding.

Before you teach

- How did children get on in the previous lesson?
- What is children's prior knowledge of column addition?
- How will you deal with misconceptions?

NATIONAL CURRICULUM LINKS

Year 4 Number – addition and subtraction

Add and subtract numbers with up to four digits using the formal written methods of columnar addition and subtraction where appropriate.

ASSESSING MASTERY

Children can use the column method to calculate. They can explain their working clearly, and understand fully what they are doing when using this method.

COMMON MISCONCEPTIONS

Children may think that they are simply adding the digits (rather than 10s, 100s etc.). Ask:
- *In 2,323 + 7,111, what does each digit represent?*

Children may not understand the importance of layout, and so may not align the columns correctly. Ask:
- *Why is it important to lay out your work correctly?*

STRENGTHENING UNDERSTANDING

To strengthen understanding, ask children to represent the numbers with base 10 equipment. This will give them a more concrete understanding.

GOING DEEPER

Deepen learning by providing children with some column additions with mistakes. Ask: *Can you spot the mistakes? Can you explain why the mistakes have been made?*

KEY LANGUAGE

In lesson: total, ones (1s), tens (10s), hundreds (100s), thousands (1,000s), add, place value, digit, altogether, addition, column

STRUCTURES AND REPRESENTATIONS

Place value grid, number line

RESOURCES

Mandatory: base 10 equipment, place value counters

 In the eTextbook of this lesson, you will find interactive links to a selection of teaching tools.

Quick recap

Write a list of 3-digit numbers on the board, arranged at random:

123; 122; 231; 331; 302; 132; 223; 201

Ask children to choose two numbers to add together and find the total. How many different totals can they find?

Discover

Unit 3: Addition and subtraction, Lesson 2

Add two 4-digit numbers

WAYS OF WORKING Pair work

ASK

- Question **1** a): *What does each digit mean in these two numbers?*
- Question **1** b): *What methods could you use?*
- Question **1** b): *How do you know you are correct?*

IN FOCUS For question **1** b), observe the different methods that children use. Many will partition the numbers and use the expanded method (which they learnt in the previous year).

PRACTICAL TIPS For this activity, leave blank place value grids on the tables for children to use if they wish.

ANSWERS

Question **1** a):

Th	H	T	O
(1,000)(1,000)(1,000)(1,000)	(100)(100)(100)(100)(100)	(10)(10)	(1)(1)(1)
(1,000)(1,000)(1,000)	(100)(100)(100)(100)	(10)(10)(10)	(1)

Question **1** b): 4,523 + 3,431 = 7,954. The two bags weigh 7,954 g in total.

Discover

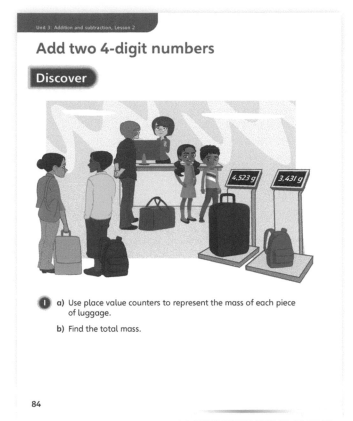

1 a) Use place value counters to represent the mass of each piece of luggage.

b) Find the total mass.

84

PUPIL TEXTBOOK 4A PAGE 84

Share

WAYS OF WORKING Whole class teacher led

ASK

- Question **1** a): *Which column shows the 1,000s? The 100s? The 10s? The 1s?*
- Question **1** b): *Do you know what to call this method of addition?*
- Question **1** b): *What sign has been used in the answer? Why?*

IN FOCUS For question **1** b), the column addition is broken down into steps, which is very important for children to see. Discuss the steps and explain them. You may want to ask children if they can think why they do not start with the thousands.

Share

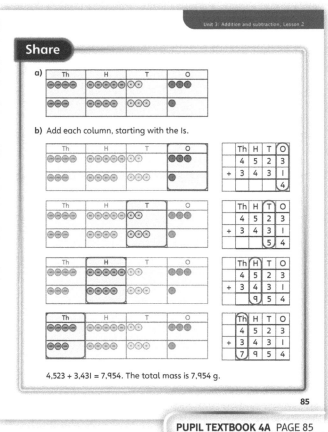

a)

b) Add each column, starting with the 1s.

4,523 + 3,431 = 7,954. The total mass is 7,954 g.

85

PUPIL TEXTBOOK 4A PAGE 85

Think together

Think together

WAYS OF WORKING Whole class teacher led (I do, We do, You do)

ASK
- Question ❶: *Which column do you add first?*
- Question ❷: *Why is it important to lay out your work correctly?*
- Question ❷: *What can you use to explain your answer?*
- Question ❸: *How can you work out the missing numbers?*

IN FOCUS Question ❸ has a column addition with missing numbers. Talk to children about how they might solve it. They will soon realise that they must do a subtraction to find the correct answer.

STRENGTHEN Provide base 10 equipment for children who need it. Children progressing from place value grids to written columns may need support with labelling the columns.

DEEPEN Deepen learning by challenging children to become the teacher. Show them a list of additions, some of which have mistakes, for example 4,556 + 2,002 = 6,008. Challenge them to mark the additions, correct the mistakes and explain where the person who made the error may have gone wrong.

ASSESSMENT CHECKPOINT Use question ❷ to see if children understand the importance of laying out their work correctly. This will also give you an insight into their understanding of place value.

ANSWERS

Question ❶: 3,142 + 2,306 = 5,448

The two bags weigh 5,448 g in total.

Question ❷: Kate has lined up the 3 hundreds under the thousands column, then the tens and ones under the hundreds and tens columns.

	Th	H	T	O
	4	5	2	1
+		3	4	6
	4	8	6	7

4,521 + 346 = 4,867

Question ❸: The mass of the second suitcase is 4,325 kg.

	Th	H	T	O
	3	4	5	2
+	**4**	**3**	**2**	**5**
	7	7	7	7

3,452 + 4,325 = 7,777

PUPIL TEXTBOOK 4A PAGE 86

PUPIL TEXTBOOK 4A PAGE 87

120

Practice

WAYS OF WORKING Independent thinking

IN FOCUS Question ❶ uses column addition in the context of money. Explain to children that this is still the same method, but ensure they understand the importance of the units.

STRENGTHEN If children are struggling with the method, work through some more additions with them, linking the column method to place value grids, base 10 equipment or both.

DEEPEN Give children some word problems in which they need to add 4-digit numbers.

THINK DIFFERENTLY In question ❸, children have to spot the mistakes. If they are struggling, prompt them to look carefully at the layout of the column additions and the numbers that are being used.

ASSESSMENT CHECKPOINT Question ❼ will allow you to assess which children have achieved mastery in this lesson. Those who have will be able to find multiple solutions using a mixture of place value knowledge and mental calculations.

ANSWERS Answers for the **Practice** part of the lesson can be found in the *Power Maths* online subscription.

Reflect

WAYS OF WORKING Pair work

IN FOCUS This activity is an excellent opportunity for children to show their understanding by teaching this topic themselves. This reinforces the saying that by learning you will teach, and by teaching you will learn.

ASSESSMENT CHECKPOINT Assess whether children can articulate their answers in simple steps, using the correct vocabulary.

ANSWERS Answers for the **Reflect** part of the lesson can be found in the *Power Maths* online subscription.

After the lesson

- Do you need to run any intervention activities to give some children a boost?
- Which children depended heavily on using equipment or place value grids?
- Was the layout that children used in their books neat and aligned?

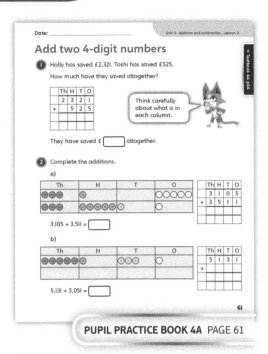

PUPIL PRACTICE BOOK 4A PAGE 61

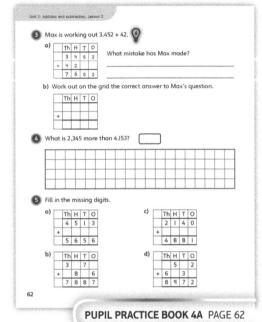

PUPIL PRACTICE BOOK 4A PAGE 62

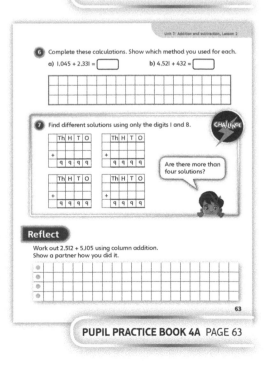

PUPIL PRACTICE BOOK 4A PAGE 63

Add two 4-digit numbers – one exchange

Learning focus

In this lesson, children will add 4-digit numbers using the column method with an exchange in one column.

Before you teach

- How will you explain the word 'exchange'?
- Would a display showing diagrams similar to those in the **Share** section of the **Textbook** support children?

NATIONAL CURRICULUM LINKS

Year 4 Number – addition and subtraction

Add and subtract numbers with up to four digits using the formal written methods of columnar addition and subtraction where appropriate.

ASSESSING MASTERY

Children can use the column method to calculate with one exchange. They can explain their working clearly and understand what the exchange means, and they can use their methods in context to solve problems.

COMMON MISCONCEPTIONS

Children often make mistakes with an exchange, for example if a column is 7 + 6, they just put the total as 0, or just put 3 without exchanging. Ask:
- *Can you explain your method to me?*

Children often do not understand that they are exchanging 10 or 100 etc., but instead think of it as 1. Ask:
- *Can you show me the exchange using place value counters?*

STRENGTHENING UNDERSTANDING

Variation in the types of addition (find the total, or add on more) will strengthen learning in this lesson. Model the exchanges with place value counters, to ensure children know what an exchange looks like.

GOING DEEPER

Use missing digit problems to deepen learning in this lesson. Can children identify what information they have and how they can use it to find missing information? Do they understand what to do when exchanges occur with missing digits?

KEY LANGUAGE

In lesson: strategy, total, addition, exchange, ones (1s), tens (10s), hundreds (100s), story problem, altogether, column method, digits

Other language to be used by the teacher: place value, thousands (1,000s)

STRUCTURES AND REPRESENTATIONS

Place value grid

RESOURCES

Mandatory: base 10 equipment, place value counters

 In the eTextbook of this lesson, you will find interactive links to a selection of teaching tools.

Quick recap

Write a list of 4-digit numbers on the board, arranged at random:

1,235; 1,225; 2,315; 3,315; 3,025; 1,325; 2,235; 2,015

Ask children to choose two numbers to add together and find the total. How many different totals can they find?

Discover

WAYS OF WORKING Pair work

ASK

- Question ❶ a): *What do the 1s add up to?*
- Question ❶ b): *What is the question asking you to do?*
- Question ❶ b): *Can you write down the number sentence?*

IN FOCUS When children represent the numbers with place value equipment in question ❶ a), they may notice that the two digits in the ones column have a total that is greater than 10. This will be significant when they come to find the total of the two numbers in question ❶ b) and an exchange is needed.

PRACTICAL TIPS For this activity, ask children to draw out and discuss the key words in the problem.

ANSWERS

Question ❶ a):

Th	H	T	O
⬤	⬤⬤⬤⬤⬤	⬤⬤⬤⬤⬤	⬤⬤⬤⬤
⬤⬤⬤⬤	⬤⬤	⬤⬤⬤	⬤⬤⬤⬤⬤⬤⬤

The 1s add up to more than 10.

Question ❶ b): 4,237 + 1,554 = 5,791. The plane will fly 5,791 miles in total.

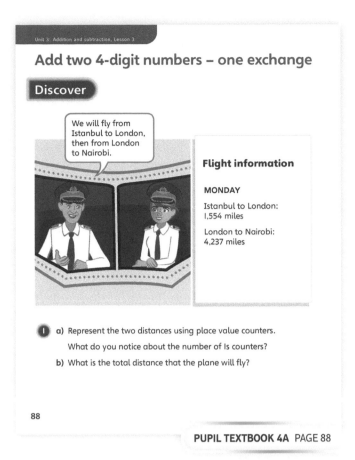

Add two 4-digit numbers – one exchange

Discover

We will fly from Istanbul to London, then from London to Nairobi.

Flight information

MONDAY

Istanbul to London: 1,554 miles

London to Nairobi: 4,237 miles

❶ a) Represent the two distances using place value counters. What do you notice about the number of 1s counters?

b) What is the total distance that the plane will fly?

88

PUPIL TEXTBOOK 4A PAGE 88

Share

WAYS OF WORKING Whole class teacher led

ASK

- Question ❶ b): *Why can you not have 11 counters in the ones column?*
- Question ❶ b): *What does 'exchange' mean?*

IN FOCUS In question ❶ a), the place value grid is used to support children in visualising the value of each digit, before they add the two numbers together. For question ❶ b), discuss why it is not possible to have 11 counters in the ones column. Show children clearly how the 10 ones are exchanged for 1 ten.

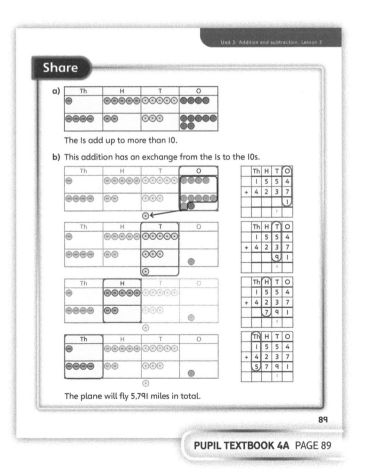

PUPIL TEXTBOOK 4A PAGE 89

Think together

WAYS OF WORKING Whole class teacher led (I do, We do, You do)

ASK

- Question **3** a): *Can you solve all of the additions mentally?*
- Question **3** a): *How do you know when you need to exchange?*
- Question **3** b): *Can you check a partner's story problems?*

IN FOCUS Question **3** a) gives an opportunity to draw out children's reasoning about when an exchange is necessary. Discuss how they can tell that an exchange is needed.

STRENGTHEN Some children may need help with layout (especially when writing the '1' when exchanging). Other children may forget to count the exchange, so give plenty of practice of this.

DEEPEN For question **3** b), deepen learning by asking: *Can you add an extra line to your story problem to make it into a multi-step problem?*

ASSESSMENT CHECKPOINT Use question **3** b) to see if children can create addition problems using 4-digit numbers with an exchange.

ANSWERS

Question **1**: 5,791 + 1,154 = 6,945
It flies 6,945 miles on Tuesday.

Question **2**: Accept any answer with an 8 or 9 in the hundreds column.

Question **3** a): Exchange 10 tens: 2,341 + 1,593 = 3,934
No exchange needed: 1,010 + 2,549 = 3,559
Exchange 10 hundreds: 7,699 = 6,917 + 782
Exchange 10 ones: 2,010 = 2,001 + 9

Question **3** b): Check that the children's story problems are appropriate and correct.

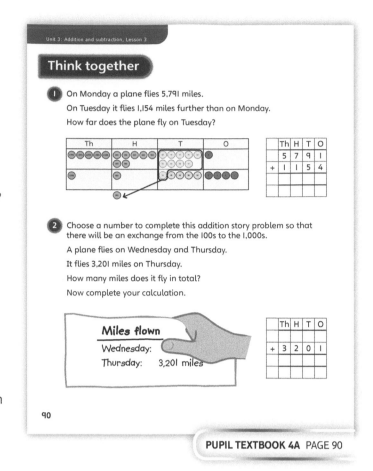

PUPIL TEXTBOOK 4A PAGE 90

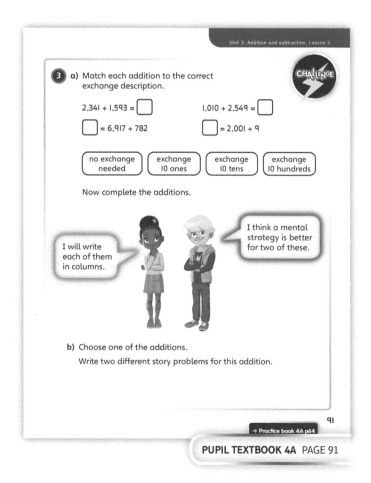

PUPIL TEXTBOOK 4A PAGE 91

Practice

WAYS OF WORKING Independent thinking

IN FOCUS You will notice that in this section there are place value counters and grids to support at the start, but not later on. This is to gradually reduce scaffolding, encouraging children to become more independent with solving additions.

STRENGTHEN If children are struggling with using the vertical column method, provide more opportunities to practise additions with an exchange.

DEEPEN Challenge children to think of number sentences in which the ones, tens and hundreds all require an exchange, for example 1,345 + 1,886. Ask them to investigate the solutions that they will get.

Give children an answer, such as 4,533. Tell them to create a question with that answer, which involves only one exchange, for example 2,822 + 1,711.

THINK DIFFERENTLY In question ④, children work back from a given total to identify missing numbers in additions where one exchange will be included.

Children may use inverse operations to find the missing digits or may complete the problems by thinking about the exchanges involved. Ask: *How can you find the missing digits? Has there been an exchange? How do you know?*

ASSESSMENT CHECKPOINT Question ④ will allow you to assess which children can complete a problem by thinking about the exchanges involved. Children who can do this confidently are likely to have mastered the lesson.

ANSWERS Answers for the **Practice** part of the lesson can be found in the *Power Maths* online subscription.

Reflect

WAYS OF WORKING Pair work

IN FOCUS In this section, children will have to think carefully about the numbers they choose and the exchanges that will happen.

ASSESSMENT CHECKPOINT Assess whether children can reason why they chose the numbers they did. You may hear comments such as, *'I needed to create an addition with an exchange of ones, so I used 9 and 2 in this column.'*

ANSWERS Answers for the **Reflect** part of the lesson can be found in the *Power Maths* online subscription.

After the lesson ⏸

- Can children explain what an exchange is?
- Can children represent an exchange with place value counters?
- Did any misconceptions crop up in this lesson?

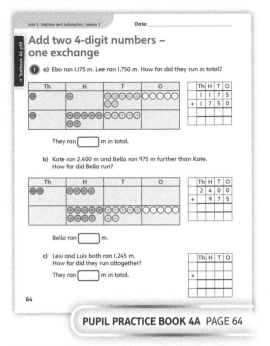

PUPIL PRACTICE BOOK 4A PAGE 64

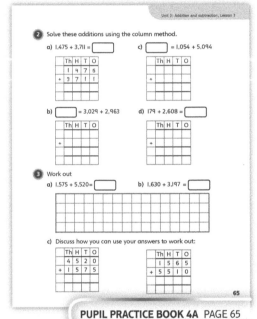

PUPIL PRACTICE BOOK 4A PAGE 65

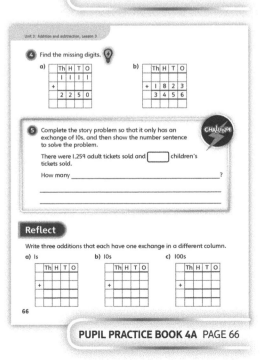

PUPIL PRACTICE BOOK 4A PAGE 66

Add with more than one exchange

Learning focus

In this lesson, children will add 4-digit numbers using the column method with exchanges across more than one column.

Before you teach

- How will you introduce calculations with multiple exchanges?
- Have you got place value counters to support children whose understanding needs strengthening?

NATIONAL CURRICULUM LINKS

Year 4 Number – addition and subtraction

Add and subtract numbers with up to four digits using the formal written methods of columnar addition and subtraction where appropriate.

ASSESSING MASTERY

Children can use the column method to calculate with more than one exchange. Children can correctly explain the methods they use, and can identify calculations that would be more suited to mental methods.

COMMON MISCONCEPTIONS

Children often make mistakes when there are multiple exchanges, for example forgetting to do the second exchange. Ask:
- *Did you remember to count the exchange?*

Children sometimes forget to add on the exchange (especially when there are multiple exchanges). Ask:
- *How could you check your answer?*

STRENGTHENING UNDERSTANDING

Together, use place value counters to work through and solve several calculations with multiple exchanges. Use the counters to make each exchange and discuss what effect it will have on the next column and on the total.

GOING DEEPER

Look at examples where one exchange leads to another exchange in the next column, such as 189 + 13. Can children mentally spot calculations where this will happen?

KEY LANGUAGE

In lesson: total, exchange, addition, ones (1s), tens (10s), hundreds (100s), method, digit, columns

Other language to be used by the teacher: place value, thousands (1,000s)

STRUCTURES AND REPRESENTATIONS

Place value grid, number line

RESOURCES

Mandatory: base 10 equipment, place value counters

 In the eTextbook of this lesson, you will find interactive links to a selection of teaching tools.

Quick recap

Write a list of 3-digit numbers on the board, arranged at random:

334; 433; 444; 343; 443; 344; 333

Ask children to choose three numbers from the list and find the total.

Discover

Add with more than one exchange

Discover

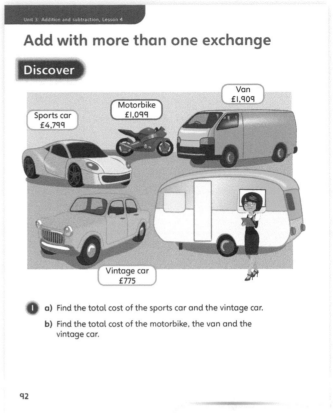

WAYS OF WORKING Pair work

ASK

• Question ① a): *What happens when there is more than one exchange in a calculation?*
• Question ① b): *Do you have to write down the calculation to see the exchanges? Can you tell by looking at the digits in each number?*

IN FOCUS In questions ① a) and ① b), focus on children's explanations of how to add with more than one exchange. Listen carefully to their vocabulary and reasoning skills.

PRACTICAL TIPS For this activity, ask children to write the calculations neatly and to highlight each of the exchanges using a different colour.

ANSWERS

Question ① a): 4,799 + 775 = 5,574. The total cost of the sports car and the vintage car is £5,574.

Question ① b): 1,099 + 1,909 + 775 = 3,783. The total cost of the motorbike, the van and the vintage car is £3,783.

① a) Find the total cost of the sports car and the vintage car.

b) Find the total cost of the motorbike, the van and the vintage car.

92

PUPIL TEXTBOOK 4A PAGE 92

Share

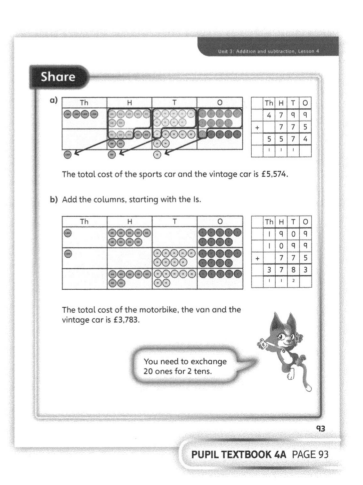

WAYS OF WORKING Whole class teacher led

ASK

• Question ① a): *Can you explain this working to me?*
• Question ① b): *How can you spot where there will be an exchange?*

IN FOCUS In question ① b), children are adding three numbers, with exchanges in several columns. They should notice that the total of the ones digits will require them to exchange two tens.

PUPIL TEXTBOOK 4A PAGE 93

Think together

WAYS OF WORKING Whole class teacher led (I do, We do, You do)

ASK

- Question **1**: *How will you lay out your work correctly?*
- Question **3** a): *Can you check your partner's additions?*
- Question **3** b): *What do you notice about each digit?*

IN FOCUS Question **3** gives children the opportunity to apply what they know about exchanges to find missing digits. They should observe what effect an exchange from one column has on the total of the column to the left of it.

STRENGTHEN Give children place value equipment to help them explain their workings. Children may need support in laying out calculations correctly on squared paper. If necessary, you could label the ones, tens, hundreds and thousands columns for them before they start their working.

DEEPEN Deepen learning in this section by providing children with a range of missing number calculations where they must decide whether or not an exchange has occurred.

ASSESSMENT CHECKPOINT Use question **3** to see whether children can find missing digits in numbers in calculations where more than one exchange has taken place and the total has 0 as a placeholder in several columns.

ANSWERS

Question **1**: 1,909 + 775 = 2,684
The van and the vintage car cost £2,684 in total.

Question **2** a): 1,099 + 1,775 = 2,874
The caravan costs £2,874.

Question **2** b): 2,874 + 1,099 = 3,973.
The caravan and motorbike cost £3,973 altogether.

Question **3** a): 1,259 + **741** = 2,000
4,**061** + 939 = 5,000
633 + 2,**36**7 = 3,000
188 + 8,8**12** = 9,000

Question **3** b): 2,716 + **284** = 3,000
9,528 + **472** = 10,000
2,104 + **7,896** = 10,000

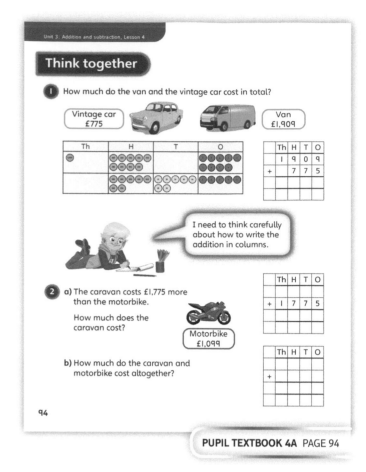

PUPIL TEXTBOOK 4A PAGE 94

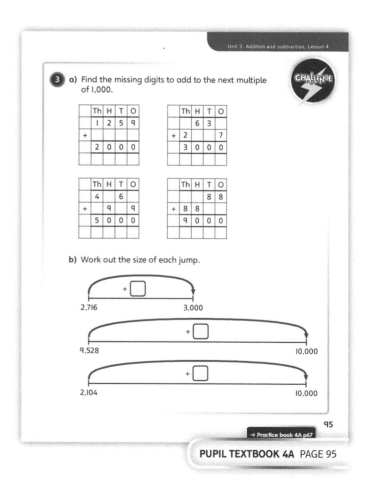

PUPIL TEXTBOOK 4A PAGE 95

Practice

WAYS OF WORKING Independent thinking

IN FOCUS Question ❹ will focus children's learning on finding calculations that have two exchanges, which should lead to mastery. Encourage children to explain their thinking verbally.

STRENGTHEN Question ❸ gives the opportunity to strengthen learning through the description of methods. Provide children who need it with questions like this but with errors where they can address examples of common misconceptions.

DEEPEN Challenge children to solve an addition with exchanges in all four columns, including the thousands. Discuss how sometimes adding two 4-digit numbers can lead to a 5-digit answer, and model the need for a ten thousands column. Ask: *Can you give me an example of an addition where one exchange causes there to be an exchange in the next column? How does the carry digit affect the column total? Why does this happen?*

Encourage children to refer to bonds to 10 in their explanation.

ASSESSMENT CHECKPOINT Question ❷ will allow you to assess whether children can solve additions that have exchanges in more than one column, or where an exchange in one column will then create an exchange in the column next to it.

ANSWERS Answers for the **Practice** part of the lesson can be found in the *Power Maths* online subscription.

Reflect

WAYS OF WORKING Independent thinking

IN FOCUS This section requires children to apply what they know about exchanges to generate an addition where several exchanges will be needed.

ASSESSMENT CHECKPOINT Assess whether children have remembered all of the steps of column addition with exchanges. You may want to encourage children to explain where the exchanges are in their calculation and why they occur.

ANSWERS Answers for the **Reflect** part of the lesson can be found in the *Power Maths* online subscription.

After the lesson ⏸

- Can children identify exchanges without having to do the workings?
- Can children represent an exchange with place value counters?
- Which children will need intervention following this lesson?

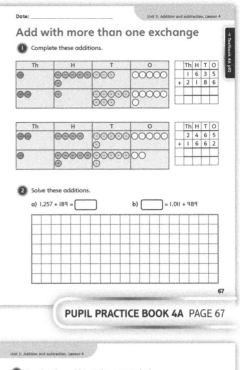

PUPIL PRACTICE BOOK 4A PAGE 67

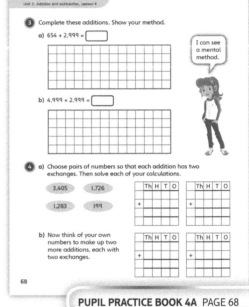

PUPIL PRACTICE BOOK 4A PAGE 68

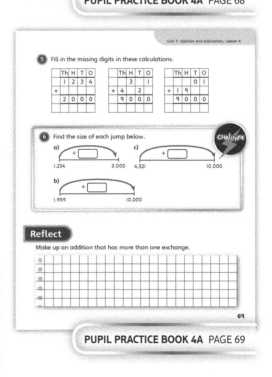

PUPIL PRACTICE BOOK 4A PAGE 69

Subtract two 4-digit numbers

Learning focus

In this lesson, children will subtract 4-digit numbers using the column method where there are no exchanges.

Before you teach

- How will you explain the key vocabulary?
- Which children are likely to struggle with the concept of place value?

NATIONAL CURRICULUM LINKS

Year 4 Number – addition and subtraction

Add and subtract numbers with up to four digits using the formal written methods of columnar addition and subtraction where appropriate.

ASSESSING MASTERY

Children can use the column method to subtract. They can explain their method clearly and demonstrate a clear understanding of place value, i.e. they know that they are not just subtracting separate digits, but recognise the 1s, 10s, 100s and 1,000s.

COMMON MISCONCEPTIONS

Children may not correctly align the columns to show 1s, 10s, 100s and 1,000s. Ask:
- *What happens if the columns are not lined up neatly?*

Some children may just subtract the digits without showing an understanding of the place value of each column. Ask:
- *Does that column show 3 – 1 or 30 – 10?*

STRENGTHENING UNDERSTANDING

As well as using place value counters to make learning more concrete, model for children how to work with bar models to represent subtractions.

GOING DEEPER

Give children a column subtraction with some digits missing and discuss what strategies they can use to complete it. Explore how to check a subtraction by finding the inverse, i.e. doing an addition.

KEY LANGUAGE

In lesson: bar model, subtraction, fewer, more than, column, digits, odd, even, story problem

Other language to be used by the teacher: place value, thousands (1,000s), hundreds (100s), tens (10s), ones (1s)

STRUCTURES AND REPRESENTATIONS

Place value grid, bar model, number line

RESOURCES

Mandatory: base 10 equipment, place value counters, dice

Optional: strips of paper to make bar models

 In the eTextbook of this lesson, you will find interactive links to a selection of teaching tools.

Quick recap

Ask children to roll three dice and use them to generate a 3-digit number. They subtract this number from 999.

Repeat, rolling the dice several more times.

Discover

Subtract two 4-digit numbers

Discover

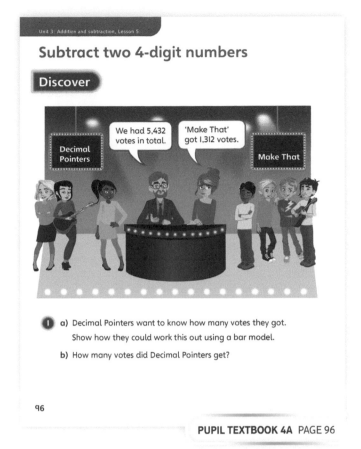

WAYS OF WORKING Pair work

ASK

- Question ① a): *How can you draw the bar model?*
- Question ① b): *How can you use the bar model to help you work out the answer?*

IN FOCUS The bar model is a powerful tool for representing subtraction. Children can clearly see the largest amount and the amount that is taken away to leave the answer. It also helps them relate subtraction to addition.

PRACTICAL TIPS Provide children with strips of paper to create their bar models.

ANSWERS

Question ① a):

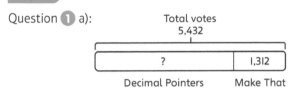

Total votes
5,432

| ? | 1,312 |

Decimal Pointers   Make That

Question ① b): 5,432 − 1,312 = 4,120
The Decimal Pointers got 4,120 votes.

① a) Decimal Pointers want to know how many votes they got. Show how they could work this out using a bar model.

b) How many votes did Decimal Pointers get?

96

PUPIL TEXTBOOK 4A PAGE 96

Share

Share

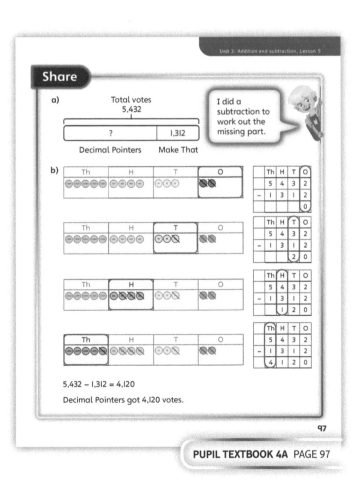

WAYS OF WORKING Whole class teacher led

ASK

- Question ① b): *How does the column method of subtraction work?*
- Question ① b): *What do the place value counters tell you about the digits in the calculation?*

IN FOCUS In question ① b), it is worth discussing that when we use place value equipment to show a subtraction, we do not need to build both numbers. We can just represent the whole, and then remove the parts that are being subtracted.

Think together

ASK

• Question ❶: *What is the question asking you to do?*
• Question ❷: *How do you work out a subtraction with a missing number?*
• Question ❸ b): *How do the models you have drawn help you to understand the subtraction?*

IN FOCUS Some children may need support with interpreting the word problems in question ❶. Use bar models to support your explanations.

STRENGTHEN Encourage children to write out and solve the subtraction in question ❷ in order to check their answer.

DEEPEN Look closely at question ❷ and discuss how to solve missing number subtractions. Draw out that a calculation like 9 – ⬜ = 5 is solved with a subtraction, i.e. 9 – 5 = 4, and a calculation like ⬜ – 4 = 5 is solved with an addition, i.e. 4 + 5 = 9.

ASSESSMENT CHECKPOINT Use question ❸ to see if children can recognise and use different representations of subtractions, which shows a deeper understanding of what a subtraction actually means.

ANSWERS

Question ❶: 4,324 – 2,120 = 2,204
　　　　　　Scissor Squares got 2,204 votes.

Question ❷: 5,465 – 264 = 5,201

Question ❸ a): 9,876 – 5,432 = 4,444
　　　　　　　9,999 – 7,654 = 2,345
　　　　　　　7,890 – 450 = 7,440

Question ❸ b): Expect children to show 7,654 – 4,321 = 3,333 using a bar model, number line and comparison bar model.

PUPIL TEXTBOOK 4A PAGE 98

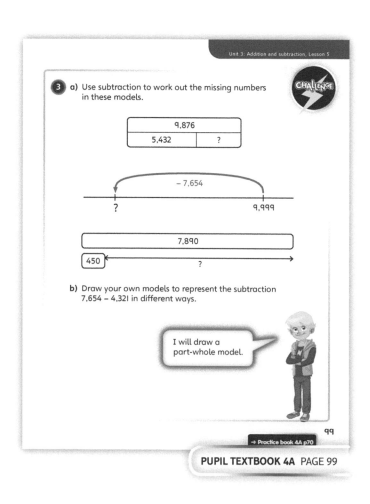

PUPIL TEXTBOOK 4A PAGE 99

132

Practice

WAYS OF WORKING Independent thinking

IN FOCUS The reasoning in question **6** will give you a good insight into which children are thinking in a deeper manner. Look for children who spot the pattern, i.e. odd – odd = even. Ask them to show you why this is.

STRENGTHEN Children may need help forming the odd and even numbers in question **6**. Make sure they understand the place value of each digit in their numbers and can see which digit is significant in making the whole number odd or even. Support them in laying their numbers out correctly for column subtraction.

DEEPEN Deepen learning in this section by giving children the answer to an unspecified subtraction. Challenge them to write down as many subtractions as they can to make that answer.

THINK DIFFERENTLY In question **5**, children have to explain why a child has made a mistake. Listen carefully to their reasoning, and prompt them to think about layout if they are finding it hard to spot the error.

ASSESSMENT CHECKPOINT Question **4** shows that children can subtract using column methods and also understand key representations.

ANSWERS Answers for the **Practice** part of the lesson can be found in the *Power Maths* online subscription.

Reflect

WAYS OF WORKING Independent thinking

IN FOCUS This activity requires children to write a story problem for a given subtraction. If necessary, guide them to stay within the context of the **Discover** section (talent show votes).

ASSESSMENT CHECKPOINT Children are likely to have mastered the lesson if they can create a problem independently and then solve it. Ask them to explain their question and solution to you.

ANSWERS Answers for the **Reflect** part of the lesson can be found in the *Power Maths* online subscription.

After the lesson ⏸

- Which children have mastered the lesson?
- Are all children ready to go on to subtractions with exchanges?
- In the next lesson what support will you provide for those children whose understanding still needs strengthening?

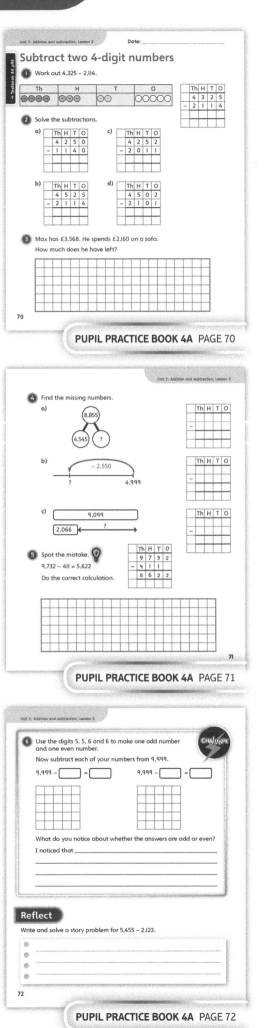

PUPIL PRACTICE BOOK 4A PAGE 70

PUPIL PRACTICE BOOK 4A PAGE 71

PUPIL PRACTICE BOOK 4A PAGE 72

Subtract two 4-digit numbers – one exchange

Learning focus

In this lesson, children will subtract 4-digit numbers using the column method where an exchange is required.

Before you teach

- How will you explain what an exchange is?
- How will you visually represent an exchange?
- Could you put something on your working wall to support children with exchanging when subtracting?

NATIONAL CURRICULUM LINKS

Year 4 Number – addition and subtraction

Add and subtract numbers with up to four digits using the formal written methods of columnar addition and subtraction where appropriate.

ASSESSING MASTERY

Children can use the column method to subtract. They can explain the method that they have used and can describe what happens when an exchange takes place (using a firm knowledge of place value).

COMMON MISCONCEPTIONS

Children may not understand how to exchange and so may say, for example, that 4 – 5 = 0 or may subtract the smaller digit from the larger, for example:

	3	4
−	2	5
	1	0

	3	4
−	2	5
	1	1

Work with children in small groups or individually to develop confidence in the method. The goal is that children feel fluent and confident with the numerical method, realising that it is more efficient to leave the counters or equipment out of the working.

STRENGTHENING UNDERSTANDING

Use place value counters and place value grids to model every exchange. Make the learning visual so that the process of exchanging is clear to understand. Run some more interventions enabling children to practise subtractions with exchanges.

GOING DEEPER

Deepen learning in this lesson by giving children a range of subtractions and asking them to work the subtractions out using more than one method. Ask: *Which method is more effective? Why?*

KEY LANGUAGE

In lesson: tens (10s), hundreds (100s), thousands (1,000s), whole, part, exchange, more, difference, method, column subtraction, number line

Other language to be used by the teacher: place value, digits, ones (1s)

STRUCTURES AND REPRESENTATIONS

Place value grid, bar model

RESOURCES

Mandatory: base 10 equipment, place value counters

Optional: string, sticky notes

In the eTextbook of this lesson, you will find interactive links to a selection of teaching tools.

Quick recap

Write a column subtraction on the board. Erase one of the digits or cover it up with a sticky note.

Challenge children to work out what the missing digit is.

Discover

ASK

- Question ① a): *Look carefully at the hundreds column. Can you see the mistake now?*
- Question ① b): *How should you lay out the subtraction?*

IN FOCUS Children are required to work out a 4-digit number minus a 3-digit number. Draw attention to the hundreds (H) column. Ask: *Why has Aki ended up with a 2 in his answer?* (He has subtracted the wrong digit.)

PRACTICAL TIPS Make a visual display in your classroom to support learning. A subtraction with an accompanying place value grid (with place value counters) is a good example to use.

ANSWERS

Question ① a): In the hundreds column, Aki has subtracted 2 hundreds from 3 hundreds, but he needed to exchange from the thousands column and then subtract 3 from 12.

Question ① b): 1,250 – 320 = 930. Aki has 930 ml of orange juice left.

Subtract two 4-digit numbers – one exchange

Discover

I think I will have 1,130 ml of orange juice left.

Aki

① a) Explain the mistake that Aki has made in his calculation.

b) How much orange juice will Aki have left?

100

PUPIL TEXTBOOK 4A PAGE 100

Share

ASK

- Question ① a): *How can you tell when an exchange is needed?*
- Question ① b): *Can you explain the method in steps?*

IN FOCUS As a class, talk through the steps of the method. In particular, highlight the importance of the place value of the digits of the numbers. Ensure children understand that 1 thousand is being exchanged for 10 hundreds, so that you can do the subtraction in the hundreds (H) column.

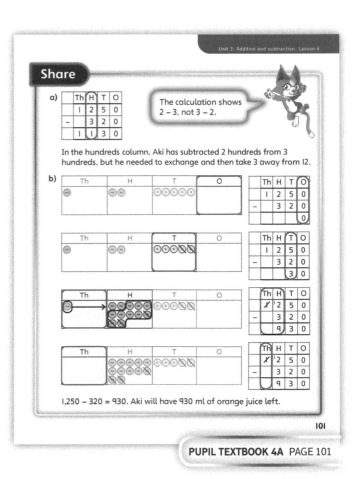

Share

a) The calculation shows 2 – 3, not 3 – 2.

In the hundreds column, Aki has subtracted 2 hundreds from 3 hundreds, but he needed to exchange and then take 3 away from 12.

b)

1,250 – 320 = 930. Aki will have 930 ml of orange juice left.

101

PUPIL TEXTBOOK 4A PAGE 101

135

Think together

WAYS OF WORKING Whole class teacher led (I do, We do, You do)

ASK

- Question **1**: *Can you tell me what the number sentence is?*
- Question **2**: *How would you show this using the column method?*
- Question **3**: *Can you use place value equipment to help you?*

IN FOCUS Questions **1** and **2** are word problems. Use a bar model to represent them, in order to help children see that they are subtractions.

STRENGTHEN For question **3**, children may need some support with identifying which column the exchange occurs in. Ask: *Can you tell me which subtraction would give the answer in each column? How can you make the numbers required?*

DEEPEN Deepen learning in this section by asking children to write their own subtractions that have one exchange. Ask: *Can you describe how you know that an exchange is needed?*

ASSESSMENT CHECKPOINT Use question **3** to assess whether children have a strong understanding of the column method of subtraction when there is one exchange.

ANSWERS

Question **1**: 1,750 − 625 = 1,125
Aki spilled 1,125 ml of mango juice.

Question **2**: 1,725 − 1,175 = 550
Lee scored 550 more points than Mo.

Question **3**: 3,253 − 2,1**26** = 1,12**7**
6,729 − **3,396** = 3,333
4,236 − 2,**723** = 1,5**13**
8,565 − 2,127 = 6,438

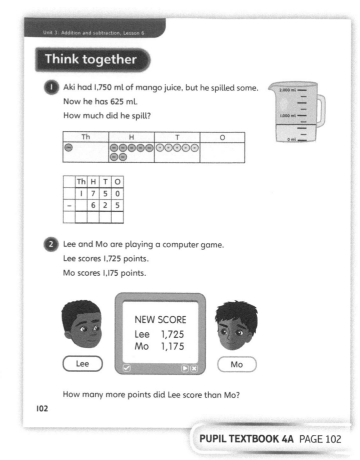

PUPIL TEXTBOOK 4A PAGE 102

PUPIL TEXTBOOK 4A PAGE 103

Practice

WAYS OF WORKING Independent thinking

IN FOCUS Question ① provides structured support with the place value grid to scaffold learning. Encourage children to describe what is happening in each column as they work through the subtractions.

STRENGTHEN Children may need some support in question ⑤. Work with them in using the information given in each calculation to work out the missing digits. Ask: *Can you identify the columns that have an exchange?*

DEEPEN Deepen learning by asking: *Can you spot where there will be an exchange without doing a calculation? How do you know? Can you write a rule for this?*

THINK DIFFERENTLY Question ④ promotes mastery of the lesson. Children will need to find the missing numbers, which is particularly challenging when there is an exchange needed. Encourage children to reason what each number must be (use place value equipment for those who require it).

ASSESSMENT CHECKPOINT Question ④ will give you an indication of which children have mastered the lesson. To assess further, ask children to talk you through their answers.

ANSWERS Answers for the **Practice** part of the lesson can be found in the *Power Maths* online subscription.

Reflect

WAYS OF WORKING Independent thinking

IN FOCUS For this exercise, put some key vocabulary on the board to support reasoning: subtraction, subtract, exchange, tens, hundreds.

ASSESSMENT CHECKPOINT Listen carefully to children's explanations. Have they identified the correct columns where the exchange is needed? Are they using correct vocabulary? Can they explain why an exchange is needed?

ANSWERS Answers for the **Reflect** part of the lesson can be found in the *Power Maths* online subscription.

After the lesson ⏸

- Are any children still making the same mistakes that they were at the start of the lesson?
- How will you tackle these misconceptions?
- Which children mastered the lesson?

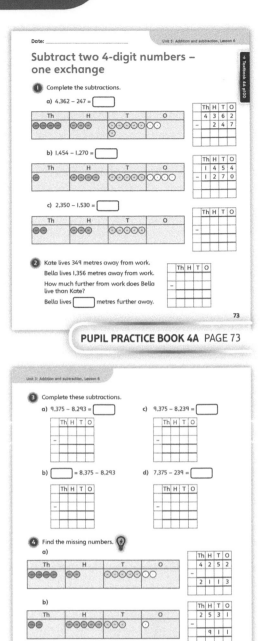

PUPIL PRACTICE BOOK 4A PAGE 73

PUPIL PRACTICE BOOK 4A PAGE 74

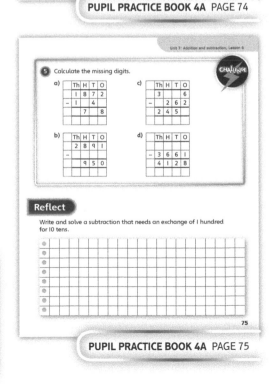

PUPIL PRACTICE BOOK 4A PAGE 75

Subtract two 4-digit numbers – more than one exchange

Learning focus

In this lesson, children will subtract 4-digit numbers using the column method where more than one exchange is required.

Before you teach

- How can you use the bar model to represent subtractions in this lesson?
- Would some children benefit from having place value counters that they can physically move to represent an exchange?

NATIONAL CURRICULUM LINKS

Year 4 Number – addition and subtraction

Add and subtract numbers with up to four digits using the formal written methods of columnar addition and subtraction where appropriate.

ASSESSING MASTERY

Children can use the column method for subtraction calculations where more than one exchange is required and can explain their answers and also check them (either with another strategy or by doing the inverse operation).

COMMON MISCONCEPTIONS

Children may forget to do one of the exchanges or may subtract the whole from the part. Ask:
- *How many exchanges were in that calculation?*

STRENGTHENING UNDERSTANDING

Keep reinforcing the place value of the digits involved in an exchange so that children learn the method and also the reasoning behind it. Continue to run quick interventions where children practise subtractions with exchanges.

GOING DEEPER

Deepen learning in this lesson by providing children with more opportunities to solve word problems that involve subtractions with more than one exchange.

KEY LANGUAGE

In lesson: difference, more, fewer, subtraction, exchange

Other language to be used by the teacher: place value, digits, thousands (1,000s), hundreds (100s), tens (10s), ones (1s), whole, part

STRUCTURES AND REPRESENTATIONS

Place value grid, bar model

RESOURCES

Mandatory: base 10 equipment, place value counters, dice

 In the eTextbook of this lesson, you will find interactive links to a selection of teaching tools.

Quick recap 🔁

Ask children to roll two dice and use them to generate a 2-digit number. They subtract this number from 100.

Repeat, rolling the dice several more times.

Discover

Unit 3: Addition and subtraction, Lesson 7

WAYS OF WORKING Pair work

ASK

- Question ① a): *How do you know if an exchange will be needed?*
- Question ① b): *Could you explain your method to a friend?*

IN FOCUS In question ① b), children are asked to work out a 4-digit number minus a 3-digit number with two exchanges. Focus their learning by asking them how they know how many exchanges are needed. Some children will rely on jotting down the column method; others will be able to spot the exchanges from their knowledge of number bonds.

PRACTICAL TIPS Ask children to make place value counters from paper or cardboard that they can physically move to represent multiple exchanges.

ANSWERS

Question ① a): Jen needs two exchanges (in the hundreds and ones columns).

Question ① b): 1,450 − 849 = 601
Jen has £601 left.

Subtract two 4-digit numbers – more than one exchange

Discover

① a) Jen buys a laptop.
How many exchanges does she need to make to work out this calculation?
£1,450 − £849?

b) How much money does she have left?

104

PUPIL TEXTBOOK 4A PAGE 104

Share

WAYS OF WORKING Whole class teacher led

ASK

- Question ① b): *How can part a) help you with this?*
- Question ① b): *Have you laid the calculation out neatly to show the exchanges?*

IN FOCUS For question ① b), focus on how children can use the information about exchanges from question ① a) to help work it out.

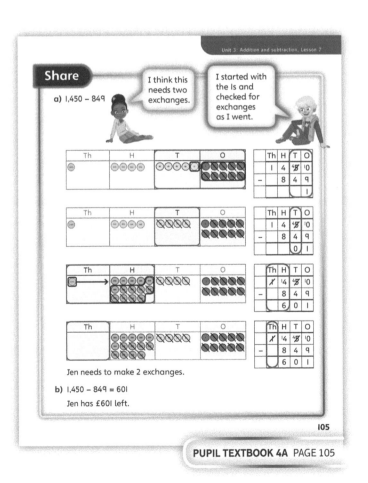

Share

a) 1,450 − 849

I think this needs two exchanges.

I started with the 1s and checked for exchanges as I went.

Jen needs to make 2 exchanges.

b) 1,450 − 849 = 601
Jen has £601 left.

105

PUPIL TEXTBOOK 4A PAGE 105

Think together

WAYS OF WORKING Whole class teacher led (I do, We do, You do)

ASK

- Question ❶: *What does 'difference' mean?*
- Question ❶: *How can you represent a difference?*
- Question ❷: *How are subtraction and addition linked?*

IN FOCUS In question ❶, the word 'difference' is in the word problem. Draw out what this term means. Stand next to a child and ask what the difference in heights is. Establish that we will need a subtraction to work it out. In question ❸, children may agree that only one exchange is needed as the digits in the hundreds and the tens columns of the part are not greater than the digits of the whole. Support them in seeing that once an exchange has been done from the 10s to the 1s, the 10s whole digit will now be smaller than the 10s part digit, and so a second exchange, from 100s to 10s, will be needed.

STRENGTHEN Question ❷ links subtraction and addition. Provide children with more subtractions to represent using a bar model. After this, challenge them to tell you the associated addition.

DEEPEN Deepen learning by asking children to think of mental strategies for the subtractions in this section. For example, in question ❷, children may suggest finding the difference by counting on in jumps from 1,880 to 2,450.

ASSESSMENT CHECKPOINT Assess children's learning by looking at their answers to question ❸. Children are likely to have mastered the lesson if they can solve the calculation and reason that one exchange caused another exchange to occur.

ANSWERS

Question ❶: 2,450 − 1,295 = 1,155
The difference in price between the two televisions is £1,155.

Question ❷: 2,450 − 1,880 = 570
The missing part is £570.

Question ❸ a): 1,295 − 199 = 1,096
The television will cost £1,096 in the sale. Astrid is not correct. Astrid will also have to exchange the tens after one of the tens has been exchanged for 10 ones.

Question ❸ b): Answers will vary. Calculations should involve at least two of: an exchange of a 10 for 1s, an exchange of a 100 for 10s or an exchange of 1,000 for 100s.

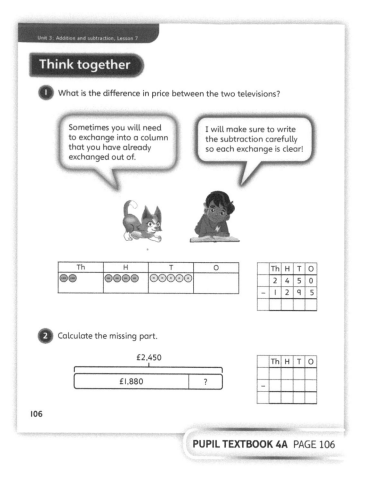

PUPIL TEXTBOOK 4A PAGE 106

PUPIL TEXTBOOK 4A PAGE 107

Practice

WAYS OF WORKING Independent thinking

IN FOCUS Question **5** provides a real challenge and will consolidate learning. Look at the subtraction, in which some of the digits are missing. If children are struggling to get started, ask them what numbers could not go in, to focus their thinking.

STRENGTHEN In question **2**, children work through a series of subtractions where more than one exchange will be needed. Ask: *Can you model the subtraction using use place value equipment? Can you identify where each exchange occurs?*

DEEPEN Ask children to make up some of their own word problems that involve multiple exchanges.

ASSESSMENT CHECKPOINT Question **6** will allow you to assess which children have mastered the lesson. Look carefully at their reasoning. Have they linked the solution to subtraction?

ANSWERS Answers for the **Practice** part of the lesson can be found in the *Power Maths* online subscription.

Reflect

WAYS OF WORKING Independent thinking

IN FOCUS This will help you to check children's understanding of the lesson. They should be able to generate a calculation where two exchanges will be needed and check the calculations of others.

ASSESSMENT CHECKPOINT Assess whether children can explain why and how exchanges occur and can use this to solve a subtraction of their choosing.

ANSWERS Answers for the **Reflect** part of the lesson can be found in the *Power Maths* online subscription.

After the lesson ⏸

- Can all children now use the column method?
- Can all children now exchange?
- Are children reasoning well?

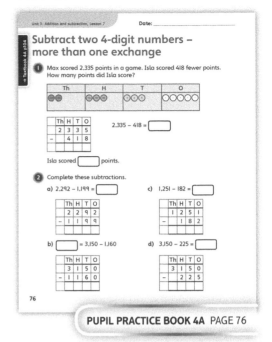

PUPIL PRACTICE BOOK 4A PAGE 76

PUPIL PRACTICE BOOK 4A PAGE 77

PUPIL PRACTICE BOOK 4A PAGE 78

Exchange across two columns

Learning focus

In this lesson, children will subtract 4-digit numbers using the column method with exchanges, when there is a zero in the column to be exchanged from.

Before you teach

- How will you explain what to do if there is a 0 in a column required for an exchange?
- Are all children ready for this lesson? Did they master the previous lesson?
- How will you support those children who did not?

NATIONAL CURRICULUM LINKS

Year 4 Number – addition and subtraction

Add and subtract numbers with up to four digits using the formal written methods of columnar addition and subtraction where appropriate.

ASSESSING MASTERY

Children can understand what to do when there is a 0 is in the column in which an exchange is required; they can talk through their methods, demonstrating a clear understanding of place value of digits in numbers. Children can also show their subtractions on a part-whole model.

COMMON MISCONCEPTIONS

When children see that there is a 0 in the next column, they may not change the 0 but still exchange '1' from it anyway. Ask:
- *What should you do if there is a 0 in the column you need for an exchange?*

Children may think that 0 minus a number is either 0 or the number itself, for example, 0 – 2 = 0 or 0 – 2 = 2. Ask:
- *Can you draw 0 – 2? Is the answer 2?*

STRENGTHENING UNDERSTANDING

Run a quick intervention in which children practise doing subtractions with a 0 in the column that is required for an exchange, for example 1,001 – 342. Encourage children to work through each step methodically.

GOING DEEPER

Deepen learning in this lesson by providing children with subtractions that have exchange mistakes in them. Can children spot the mistakes and reason why they may have been made?

KEY LANGUAGE

In lesson: subtraction, exchange, ones (1s), tens (10s), hundreds (100s), column method, zero (0), place value, partition

Other language to be used by the teacher: thousands (1,000s), whole, part

STRUCTURES AND REPRESENTATIONS

Place value grid, bar model, part-whole model

RESOURCES

Mandatory: base 10 equipment, place value counters, dice

 In the eTextbook of this lesson, you will find interactive links to a selection of teaching tools.

Quick recap 𝒬

Ask children to roll three dice and use them to generate a 3-digit number. They subtract this number from 1,000.

Repeat, rolling the dice several more times.

Discover

Pair work

ASK

- Question **1** a): *Could Bella look at the next column?*
- Question **1** a): *How could the hundreds column help with the exchange?*

IN FOCUS Children are exploring what happens when an exchange is needed but there is a 0 in the next column. Listen carefully to what they think they should do, prompting them to look at the next column if necessary.

PRACTICAL TIPS Use place value counters to make numbers with 0 in some columns. Discuss what will happen if we try to subtract from a column where there are no counters, and model exchanging from the next column.

ANSWERS

Question **1** a): Bella wants to exchange a ten for 10 ones, but she cannot because 2,502 doesn't have any tens. First, Bella should exchange 1 hundred for 10 tens. Then she can exchange 1 ten for 10 ones.

Question **1** b): 2,502 − 243 = 2,259

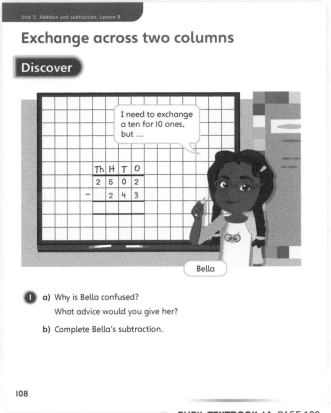

PUPIL TEXTBOOK 4A PAGE 108

Share

Whole class teacher led

ASK

- Question **1** b): *Do you understand how the exchange can be made with the 100s?*
- Question **1** b): *What steps do you need to take when there is a 0 in the column you need for an exchange?*

IN FOCUS For question **1** b), model the subtraction in the column method format. Show children the importance of the layout, how to strike through the numbers being exchanged and how to put a small 1 for the exchanged number.

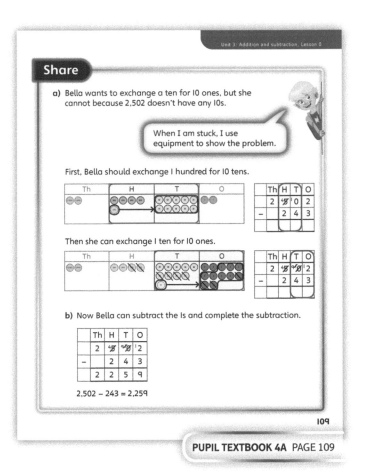

PUPIL TEXTBOOK 4A PAGE 109

Think together

Think together

WAYS OF WORKING Whole class teacher led (I do, We do, You do)

ASK

- Question ❷: *Can you use the words 'exchange', 'hundreds' and 'ones' in your explanation?*
- Question ❸: *What do you do when there is a 0 in the tens column and the hundreds column?*
- Question ❸: *How can you predict the exchanges you will need to make before you do the subtraction?*

IN FOCUS In question ❸, children are faced with 5,005 – 2,929. They will see that there is a 0 in the tens column and in the hundreds column. Explain that children should follow the same method they previously learnt, working back from the thousands column. Model an example on the board.

STRENGTHEN Question ❶ a) requires children to exchange from the 100s and the 10s in order to subtract in the ones column. Ask them to carefully describe each step aloud as they do it, to clarify what is needed. For example, *I need to exchange 1 ten but there are none. I can take 1 hundred from the hundreds column, that is 10 tens.*

DEEPEN Question ❸ links to previous work on subtracting by using knowledge of the place value of digits in numbers. Children should be able to spot the links between the calculations: for example, the hundreds and tens digits are switched in 5,055 – 2,929 and 5,505 – 2,929.

ASSESSMENT CHECKPOINT Question ❷ will show you which children are confident with subtracting when there are 0s in columns that require exchanges. Listen carefully to their explanations of where Zac went wrong, and then assess their corrected answer.

ANSWERS

Question ❶ a): 2,032 – 512 = 1,520

Question ❶ b): 5,403 – 505 = 4,898

Question ❷: Zac forgot to exchange 1 hundred for 10 tens, and 1 ten for 10 ones. He needs to cross out after exchanging, and write the new number of hundreds or tens.
3,304 – 1,269 = 2,035

Question ❸: 1 thousand for 10 hundreds and 1 ten for 10 ones: 2,126 = 5,055 – 2,929
1 thousand for 10 hundreds, 1 hundred for 10 tens and 1 ten for 10 ones: 2,576 = 5,505 – 2,929
1 thousand for 10 hundreds, 1 hundred for 10 tens and 1 ten for 10 ones: 2,076 = 5,005 – 2,929
1 thousand for 10 hundreds and 1 hundred for 10 tens: 2,480 = 5,005 – 2,525

PUPIL TEXTBOOK 4A PAGE 110

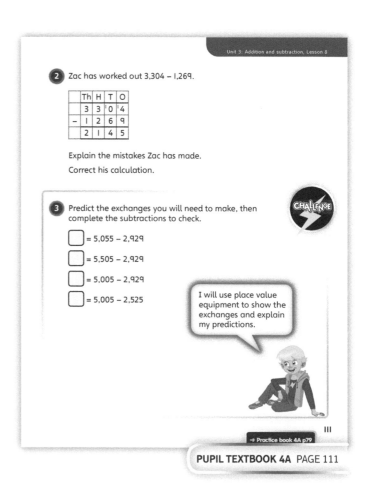

PUPIL TEXTBOOK 4A PAGE 111

Practice

WAYS OF WORKING Pair work

IN FOCUS In question **1**, children use the column method for subtractions that involve more than one exchange. Questions **2** and **3** are problem-solving questions which require subtraction with more than one exchange in a real-life money context. Question **4** requires children to identify and correct errors in the column method of subtraction. Question **5** presents children with a series of abstract calculations involving numbers that have 0 as a placeholder. In question **6**, multi-step calculations will be required.

STRENGTHEN Ask children to represent their working using place value counters. They should use and move place value counters in real life, to model each step and consolidate what happens each time you make an exchange, particularly where there is a 0 in one or more columns.

DEEPEN Question **2** provides a subtraction in the context of words and reading. Provide more examples of story problems with real-life contexts like this, in particular, problems with multiple steps which will deepen learning even further.

ASSESSMENT CHECKPOINT Use question **5** to assess whether children can correctly use the column method of subtraction where more than one exchange will be needed.

ANSWERS Answers for the **Practice** part of the lesson can be found in the *Power Maths* online subscription.

Reflect

WAYS OF WORKING Pair work

IN FOCUS Consider having a list of key vocabulary on the board or learning wall to support children's explanations.

ASSESSMENT CHECKPOINT Assess children on whether they can correctly explain their methodology, for example: *If I need to exchange 10 ones, when there is a 0 in the tens column, I must first exchange 1 hundred for 10 tens, then exchange 1 ten for 10 ones.*

ANSWERS Answers for the **Reflect** part of the lesson can be found in the *Power Maths* online subscription.

After the lesson

- Can all children explain how to exchange when there is a 0 in a column required for the exchange?
- Are children ready to start exploring efficient methods of subtraction?
- Should you run some intervention sessions?

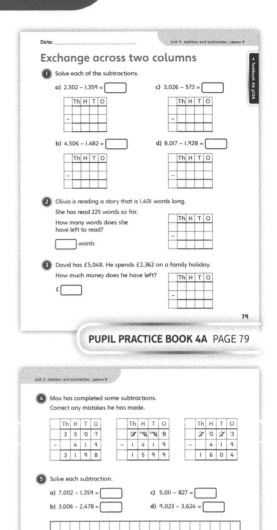

PUPIL PRACTICE BOOK 4A PAGE 79

PUPIL PRACTICE BOOK 4A PAGE 80

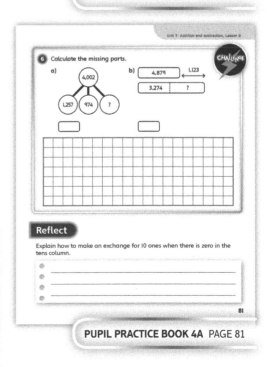

PUPIL PRACTICE BOOK 4A PAGE 81

Efficient methods

Learning focus

In this lesson, children consider different methods for solving calculations, thinking about how to work efficiently and accurately.

Before you teach

- Can children use column addition accurately?
- Can children use column subtraction accurately?
- Can children explain how and why 'exchanging' works in addition and subtraction methods?

NATIONAL CURRICULUM LINKS

Year 4 Number – number and place value

Add and subtract numbers with up to four digits using the formal written methods of columnar addition and subtraction where appropriate.

Estimate and use inverse operations to check answers to a calculation.

ASSESSING MASTERY

Children can make reasoned decisions when choosing the most appropriate calculation method to use.

COMMON MISCONCEPTIONS

Children may become confused when they are presented with multiple methods. Ask:
- *Which method do you prefer? Why?*
- *Why do you think some people choose different methods?*

STRENGTHENING UNDERSTANDING

Explain that although there may be several possible methods, the column method can always be used, if you are able follow the process accurately. Encourage children to use place value equipment to support their reasoning.

GOING DEEPER

Challenge children to evaluate a range of different methods that could be used to solve a particular calculation, for example 8,005 – 6,995. Ask: *Which method did you prefer? Why?*

KEY LANGUAGE

In lesson: column method, mental method, written method

Other language to be used by the teacher: exchange, digit

STRUCTURES AND REPRESENTATIONS

Column methods, number lines

RESOURCES

Optional: place value equipment

 In the eTextbook of this lesson, you will find interactive links to a selection of teaching tools.

Quick recap 🔁

Ask children to count on in 1,000s from a given 3-digit number.

Then ask them to count on in 100s from a given 3-digit number.

Discover

ASK

- Question ① a): *What's the same and what's different about these calculations?*
- Question ① a): *What do you notice about the second number in each calculation? How would you do each calculation if this were 1,000 instead of 999?*

IN FOCUS Children will be observing the numbers involved in given calculations and considering how this affects the thinking needed and the method that may be selected to find the answer.

PRACTICAL TIPS Before looking at the calculations in the question, ask children to locate numbers such as 9, 49, 99, 199, 999, 4,999 on number lines. Ask: *What do you notice?*

ANSWERS

Question ① a):

	Th	H	T	O
	1	5	1	7
+		9	9	9
	2	5	1	6
		1	1	1

	Th	H	T	O
	$\not{1}$	$^{14}\not{5}$	$^{10}\not{1}$	$^{1}7$
−		9	9	9
			5	1	8

Question ① b): Other possible methods include partitioning using part-whole models and counting on with number lines. 999 is very near to 1,000, so adding or subtracting 1,000 and adjusting by the extra 1 is an efficient mental method.

Share

ASK

- Question ① a): *How many exchanges are needed in each calculation?*
- Question ① a): *How many steps do you need to solve each of these column methods?*
- Question ① b): *Can you see how the different calculations are related?*
- Question ① b): *does adding or subtracting 1,000 help you to add or subtract 999?*

IN FOCUS Question ① a) requires children to use the column methods of addition and subtraction accurately, to solve calculations that will include multiple exchanges. In question ① b), children explore the use of efficient mental methods involving compensation strategies with numbers that are near multiples of 1,000.

Efficient methods

Discover

1,517 + 999
1,517 − 999

① a) Solve each calculation using a column method.

b) Discuss other possible methods to solve these calculations.

112

PUPIL TEXTBOOK 4A PAGE 112

Share

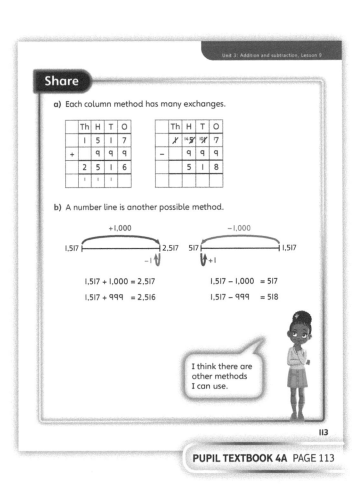

a) Each column method has many exchanges.

b) A number line is another possible method.

1,517 + 1,000 = 2,517
1,517 + 999 = 2,516

1,517 − 1,000 = 517
1,517 − 999 = 518

I think there are other methods I can use.

113

PUPIL TEXTBOOK 4A PAGE 113

Think together

WAYS OF WORKING Whole class teacher led (I do, We do, You do)

ASK

- Question ❶: *Which method would be best with these numbers?*
- Question ❷: *How can you adapt the method each time?*
- Question ❸: *What do you notice about 2,001 and 1,998? How close are they?*

IN FOCUS Question ❶ gives children the opportunity to explore the merits of the column written method and different mental strategies, and to find that they both give the same answer. The question does not teach the mental method as a trick, instead it gives children the chance to evaluate both methods themselves. You may want to model the mental method using a number line.

Question ❷ explores the compensation method with near multiples of 10, 100 or 1,000. Children either add 10, 100 or 1,000 and then subtract 1 to adjust, or they subtract 10, 100 or 1,000 and then add 1. Encourage children to spot the patterns in the calculations and to reason on which digits change and which ones stay the same. In question ❸, children use different mental methods to subtract. Encourage children to describe their methods, such as: counting on or back to 10s or 1,000s; using number bonds; using the compensation method and other efficient mental strategies.

STRENGTHEN Focus on asking children to use mental methods with calculations that are in lower number ranges, in order to build up their confidence.

DEEPEN Refer children to question ❷ and ask them to consider the mental methods that they might use to add other near multiples of 10, 100 or 1,000 such as 49, 199, 995, and so on.

ASSESSMENT CHECKPOINT Use question ❶ to see whether children can choose between a written method and a mental method. In particular, listen to their reasoning.

ANSWERS

Question ❶ a): 1,999 + 575 = 2,574
Other methods could include the column method or the compensation method, i.e. 1,999 + 575 = 2,000 + 575 − 1 = 2,575 − 1 = 2,574. Alternatively, they may reason that since 575 = 1 + 574, so 1,999 + 575 = 1,999 + 1 + 574 = 2,574.

Question ❶ b): Calculations should involve numbers one less than a multiple of a hundred or a thousand, such as 3,999 or 299.

Question ❷: 5,207 + 9 (= 5,207 + 10 − 1) = 5,216
5,207 + 99 (= 5,207 + 100 − 1) = 5,306
5,207 + 999 (= 5,207 + 1,000 − 1) = 6,206
5,207 − 9 (= 5,207 − 10 + 1) = 5,198
5,207 − 99 (= 5,207 − 100 + 1) = 5,108
5,207 − 999 (= 5,207 − 1,000 + 1) = 4,208

Question ❸ a): 2,001 − 1,998 = 3; 2,001 − 5 = 1,996

Question ❸ b): 2,991 − 2 = 2,989; 2,001 − 9 = 1,992; 1,001 − 999 = 2.

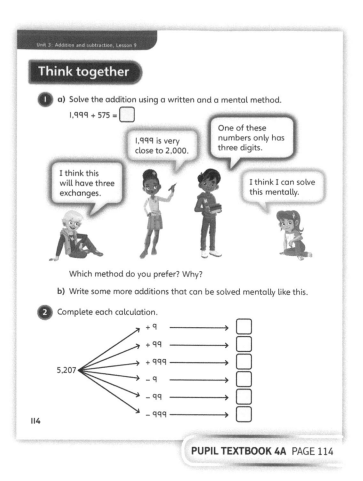

PUPIL TEXTBOOK 4A PAGE 114

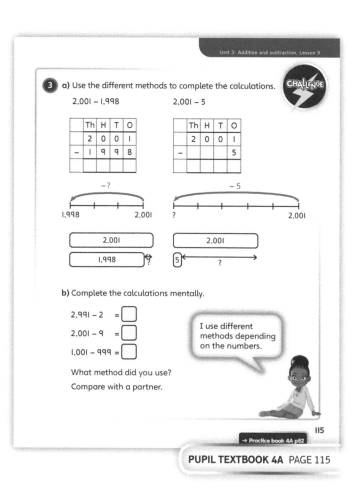

PUPIL TEXTBOOK 4A PAGE 115

Practice

WAYS OF WORKING Independent thinking

IN FOCUS In questions ❶, ❷ and ❸, children rehearse their use of compensation methods when adding or subtracting near multiples of 10, 100 or 1,000. Question ❹ requires children to demonstrate different mental methods for subtraction calculations such as: counting on or back to 10s or 1,000s; using number bonds; using the compensation method and other efficient mental strategies. In question ❺, children select the most efficient method each time.

STRENGTHEN Children may need some support in question ❹. Provide 'big' number lines (for example, string) and place value counters, so that they can physically represent the problems.

DEEPEN Challenge children to find five different methods that they could use to solve 4,111 – 989. Ask: *Which method did you prefer? Why?*

ASSESSMENT CHECKPOINT Assess whether children are able to justify their choices about the methods that they choose to use for a variety of different calculations.

ANSWERS Answers for the **Practice** part of the lesson can be found in the *Power Maths* online subscription.

Reflect

WAYS OF WORKING Independent thinking

IN FOCUS The **Reflect** part of the lesson provides children with the opportunity to make decisions about when written methods or mental methods would be more appropriate.

ASSESSMENT CHECKPOINT Assess whether children can explain why certain calculations lend themselves to written or mental methods.

ANSWERS Answers for the **Reflect** part of the lesson can be found in the *Power Maths* online subscription.

After the lesson ⏸

- Can children discuss and compare two different methods that could be used to solve 2,554 + 998?
- Can they discuss and compare two different methods that could be used to solve 2,554 – 998?

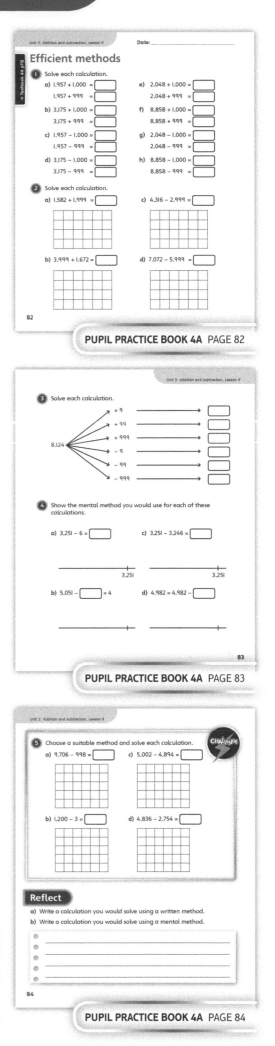

PUPIL PRACTICE BOOK 4A PAGE 82

PUPIL PRACTICE BOOK 4A PAGE 83

PUPIL PRACTICE BOOK 4A PAGE 84

Equivalent difference

Learning focus

In this lesson, children will learn the equivalent difference method of subtraction.

Before you teach

- Do children know what 'equivalent' means?
- How will you explain it?
- Are children comfortable with using the bar model to represent differences?

NATIONAL CURRICULUM LINKS

Year 4 Number – addition and subtraction

Estimate and use inverse operations to check answers to a calculation.

ASSESSING MASTERY

Children understand the equivalent difference strategy and can apply it when solving problems. Children can explain the method correctly and suggest why it is more efficient than the column method (or a mental method).

COMMON MISCONCEPTIONS

Children often do not understand the reasoning behind equivalent difference, that you can adjust the two numbers in a subtraction so that the difference remains the same. With the subtraction 232 – 98, they may know that it is easier to subtract 100 and that they will need to adjust 232 accordingly, but they are not sure whether it should be 230 – 100 or 234 – 100. Ask:
- *Can you show the equivalent difference on a bar model?*

STRENGTHENING UNDERSTANDING

For children who need more support, use a bar model to represent subtractions, drawing attention to equivalent differences. Ask children to draw their own bar models to represent each problem.

GOING DEEPER

Deepen learning in this lesson by asking children to think of more than one strategy that they can use to solve a subtraction, for example equivalent difference, column method and counting on to find the difference. Ask them to reason which method is the most efficient.

KEY LANGUAGE

In lesson: difference, subtraction, exchange, bar model, method, **efficient**, column, equivalent

Other language to be used by the teacher: place value, digits, thousands (1,000s), hundreds (100s), tens (10s), ones (1s), fewer

STRUCTURES AND REPRESENTATIONS

Bar model

RESOURCES

Place value equipment, dice

 In the eTextbook of this lesson, you will find interactive links to a selection of teaching tools.

Quick recap

Ask children to roll three dice and use them to generate a 3-digit number. They repeat this to generate a second number. They then find the difference between the two numbers.

Discover

Unit 3: Addition and subtraction, Lesson 10

Equivalent difference

WAYS OF WORKING Pair work

ASK

- Question ① b): *How can you work out the difference?*
- Question ① b): *Can you spot the related calculations that do not require an exchange?*

IN FOCUS In question ① b), children should explore an efficient method to work out the subtraction. If necessary, draw their attention to the calculations that do not require an exchange.

PRACTICAL TIPS Discuss the context of the question and consider comparing the age of a real person, perhaps a singer or celebrity, with children's ages. This will make the learning more real, and will spark enthusiasm for real-life applications of the topic.

ANSWERS

Question ① a): Amelia's statement is not true. The difference between Amelia's age and her great-grandad's age will always be the same. Children may use a variety of methods to show this. Accept any that show that the difference will not change.

Question ① b): The difference between their ages will be 88 years.

Discover

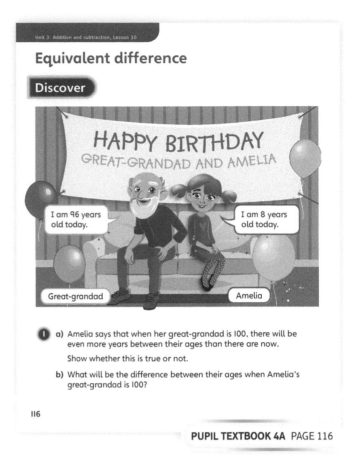

① a) Amelia says that when her great-grandad is 100, there will be even more years between their ages than there are now.

Show whether this is true or not.

b) What will be the difference between their ages when Amelia's great-grandad is 100?

116

PUPIL TEXTBOOK 4A PAGE 116

Share

WAYS OF WORKING Whole class teacher led

ASK

- Question ① a): *What do you notice about the difference when both of the ages increase?*
- Question ① b): *Which subtractions are the easiest to calculate mentally?*

IN FOCUS For question ① a), make it clear that if we increase each number by the same amount, the difference will not change.

Share

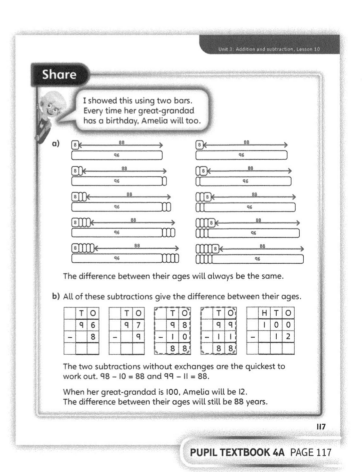

151

Think together

Whole class teacher led (I do, We do, You do)

ASK

- Question **2**: *Why did you choose that calculation to find the difference?*
- Question **3** b): *Which methods can you remember from things you have learnt before?*
- Question **3** b): *Which methods were the most efficient?*

IN FOCUS In question **2**, children should realise that two subtractions do not involve an exchange (128 − 100 and 129 −101). Then they should reason that 128 − 100 is easier to work out as they are simply subtracting 100.

STRENGTHEN Run a quick intervention for children to practise solving more subtractions where they can use equivalent difference to get to a multiple of 100 or 10, which will be more efficient, for example, 154 − 97, 222 − 198, 100 − 58.

DEEPEN Children can use question **3** to explore different strategies for working out subtractions. Encourage them to evaluate each subtraction and consider which would be the most efficient method each time.

ASSESSMENT CHECKPOINT Question **3** b) will give you an insight into which children have mastered the equivalent difference method. It will also tell you if children can apply other learnt strategies to subtractions, based on which method is more efficient.

ANSWERS

Question **1**: 198 − 79 = 119
199 − 80 = 119
200 − 81 = 119
The difference is 119 years.

Question **2**: 125 − 97 = 28
126 − 98 = 28
127 − 99 = 28
128 − 100 = 28
129 − 101 = 28
The whale is 28 years younger than the giant tortoise.

Question **3** a): 1,000 − 245 = 755
Astrid's method is more efficient because no exchanges are needed.

Question **3** b): 1,000 − 542 = 458
2,692 − 836 = 1,856
2,001 − 265 = 1,736
1,897 − 999 = 898
Check for a range of strategies being used, for example, equivalent difference, column method, counting on to find the difference (number line), counting back (number line), expanded method.

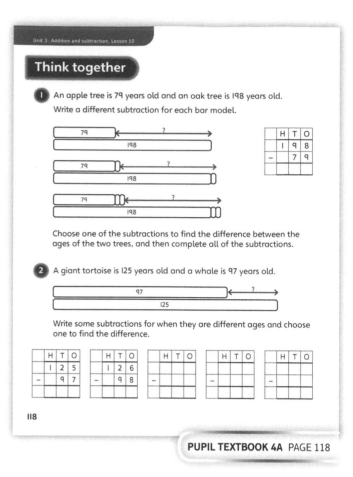

PUPIL TEXTBOOK 4A PAGE 118

PUPIL TEXTBOOK 4A PAGE 119

Practice

WAYS OF WORKING Independent thinking

IN FOCUS Question **5** is an open question in which children can choose from a range of methods that they know, identifying the most efficient. When they have finished, ask them to compare their chosen strategies with a partner. Encourage them to use correct vocabulary in their discussion.

STRENGTHEN Encourage children to show each of their methods using a representation such as a bar model or a number line.

DEEPEN Focus on question **1**. Ask children if they can think of a subtraction for which it would not be a good idea to use equivalent difference.

ASSESSMENT CHECKPOINT Question **1** will allow you to assess which children can find equivalent subtractions and select the most efficient one. Question **4** b) will allow you to assess children's reasoning skills around this method. If they are correct, then it is likely that they have achieved mastery of this lesson.

ANSWERS Answers for the **Practice** part of the lesson can be found in the *Power Maths* online subscription.

Reflect

WAYS OF WORKING Pair work

IN FOCUS For this question, children should reason that the column method would require three exchanges, so it is not a very efficient strategy for this subtraction. Children may instead opt to use equivalent difference or counting on from 955 to 1,000 to find the difference.

ASSESSMENT CHECKPOINT This exercise will allow you to assess whether children understand that different methods are more suitable for different subtractions. Their reasoning will let you know if they understand why this is the case.

ANSWERS Answers for the **Reflect** part of the lesson can be found in the *Power Maths* online subscription.

After the lesson

- Do all children understand equivalent difference?
- Can children visually represent equivalent difference?
- Do you need to run intervention sessions for any children?

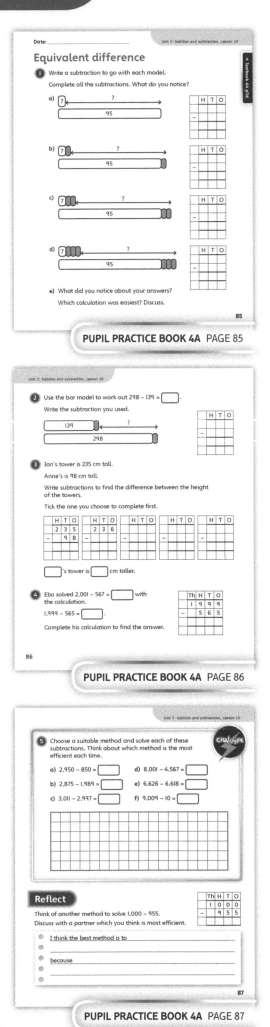

PUPIL PRACTICE BOOK 4A PAGE 85

PUPIL PRACTICE BOOK 4A PAGE 86

PUPIL PRACTICE BOOK 4A PAGE 87

Estimate answers

Learning focus

In this lesson, children will learn to make choices about whether to round to the nearest 10, 100 or 1,000 and how to use that to decide if a calculation is accurate.

Before you teach

- Can all children round to the nearest 10, 100, 1,000?
- Could you do a mini-assessment prior to the lesson?
- Would displaying the key vocabulary support children in this lesson?

NATIONAL CURRICULUM LINKS

Year 4 Number – addition and subtraction

Estimate and use inverse operations to check answers to a calculation.

ASSESSING MASTERY

Children can round the numbers in additions and subtractions up or down to the nearest 10, 100 or 1,000 as appropriate and can use this to make estimates and find rough answers. They can compare their estimates to the exact answers and use this to check answers.

COMMON MISCONCEPTIONS

Children may not know whether to round a number to the nearest 10, 100 or 1,000. Ask:
- *How accurate do you need to be? Can you work that out mentally?*

STRENGTHENING UNDERSTANDING

If children are finding rounding difficult, you may need to run some intervention for them to practise this important skill.

GOING DEEPER

This lesson shows children that sometimes it is more accurate to round only one number, to retain better accuracy. Generate discussion around this and ask children to come up with examples that demonstrate it.

KEY LANGUAGE

In lesson: accurate, exact, estimate, round, roughly, nearest, thousand, hundred, ten, one, column, subtraction, addition, check, efficient

Other language to be used by the teacher: approximately

STRUCTURES AND REPRESENTATIONS

Number line

RESOURCES

Printed number lines from 1,000 to 2,000 and from 3,000 to 4,000

 In the eTextbook of this lesson, you will find interactive links to a selection of teaching tools.

Quick recap

Ask children to add together any two given multiples of 1,000.

Then, ask them to subtract any given multiple of 1,000 from another.

Discover

WAYS OF WORKING Pair work

ASK

- Question ① a): *How can you tell the ringmistress has not made a good estimate?*
- Question ① b): *What strategy could you use to check?*

IN FOCUS The ringmistress gives an incorrect statement because she has rounded one of the amounts incorrectly. Question ① a) provides a good learning opportunity for children to spot this. This is a good point in the lesson at which to remind children of the rules involved in rounding.

PRACTICAL TIPS Provide printed number lines from 1,000 to 2,000 and from 3,000 to 4,000 for children who need visual support with the questions.

ANSWERS

Question ① a): This is not an accurate estimate. 1,898 is closer to 2,000 than 1,000. A better estimate would be 2,000 + 3,000 = 5,000. They have sold roughly 5,000 tickets.

Question ① b): The exact answer is 4,914 tickets. 4,914 rounds to 5,000. 5,000 is close to the exact calculation. 4,000 is not. 5,000 is a better estimate.

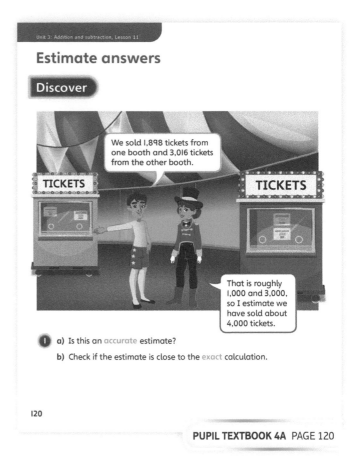

PUPIL TEXTBOOK 4A PAGE 120

Share

WAYS OF WORKING Whole class teacher led

ASK

- Question ① a): *How do the number lines help you round?*
- Question ① b): *What would the estimate be if the ringmistress had rounded to the nearest hundred?*

IN FOCUS Question ① a) gives a good opportunity to ask children what happens if the amount is exactly in the middle of the number line, i.e. 1,500. Recap the rule that they must always round up when that is the case. Explain that the five digits that round down are 0, 1, 2, 3, 4, and the digits that round up are 5, 6, 7, 8, 9. This makes it equal.

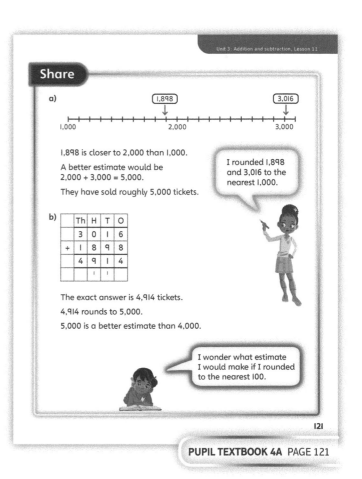

PUPIL TEXTBOOK 4A PAGE 121

Think together

ASK

- Question **2**: *What is the next 1,000 after 9,000?*
- Question **3**: *How close is Isla's estimate to Max's answer?*

IN FOCUS In question **2**, children will need to round 9,811 to the nearest 1,000. This may confuse some children who do not see that 10,000 is the answer. Model the question on a number line to provide support, if needed.

STRENGTHEN Practise counting on and back in 10s, 100s and 1,000s. This will be particularly useful in question **2**, when the 10,000s barrier is crossed.

DEEPEN Question **3** deepens learning by showing that although Isla has only rounded one of the numbers, hers is the best estimate. She can subtract a thousands number, which is a sensible mental method. Discuss this with the children and challenge them to find other subtractions in which it would be best to round only one of the numbers.

ASSESSMENT CHECKPOINT Question **2** will allow you to assess which children can round numbers and subtract them mentally to reach an estimate.

ANSWERS

Question **1**: 6,149 rounds to 6,000. 912 rounds to 1,000.
  6,000 − 1,000 = 5,000.
  Roughly 5,000 people stayed.

Question **2** a): 3,000 + 4,000 = 7,000;
  10,000 + 3,000 = 13,000

Question **2** b): 2,800 + 3,900 = 6,700;
  9,800 + 2,800 = 12,600

Question **2** c): 2,794 + 3,911 = 6,705;
  9,811 + 2,788 = 12,599
  Rounding to the nearest 100 is much more accurate but more difficult to work out mentally.

Question **3** a, b): Children should notice that Isla's method is more accurate because it is nearer to the actual answer. Although Isla has only rounded one of the numbers, her estimate is better because when you round 5,602 to the nearest thousand, you lose quite a lot of accuracy.

Question **3** c): An estimate (such as the one Isla made) would have helped Max to find out whether his answer was close to the correct one.

Think together

1 At a theatre, there were 6,149 people in the audience, but 912 of them left during the interval.

Round to the nearest 1,000 to estimate how many people stayed.

```
|----|----|----|----|----|----|----|----|----|----|
0  1,000 2,000 3,000 4,000 5,000 6,000 7,000 8,000 9,000 10,000
```

6,149 rounds to ☐,000 to the nearest 1,000.

912 rounds to ☐,000 to the nearest 1,000.

> I can add and subtract these mentally.

Approximately ☐ people stayed.

2 | 2,794 + 3,911 | | 9,811 + 2,788 |

a) Estimate the answer to each calculation by rounding to the nearest 1,000.

b) Estimate the answer to each calculation by rounding to the nearest 100.

c) Now work out the exact answers.
  What do you notice?

122

PUPIL TEXTBOOK 4A PAGE 122

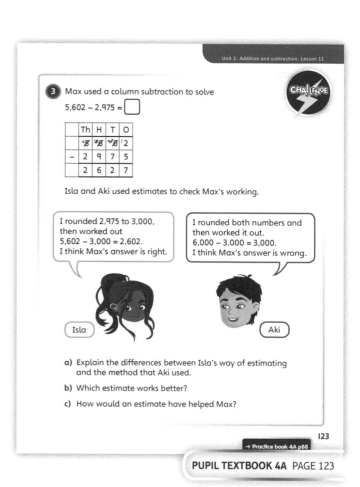

3 Max used a column subtraction to solve

5,602 − 2,975 = ☐

Th	H	T	O
⁴5̷	⁵6̷	⁹0̷	¹2
− 2	9	7	5
2	6	2	7

Isla and Aki used estimates to check Max's working.

> I rounded 2,975 to 3,000, then worked out
> 5,602 − 3,000 = 2,602.
> I think Max's answer is right.

> I rounded both numbers and then worked it out.
> 6,000 − 3,000 = 3,000.
> I think Max's answer is wrong.

Isla   Aki

a) Explain the differences between Isla's way of estimating and the method that Aki used.

b) Which estimate works better?

c) How would an estimate have helped Max?

123

→ Practice book 4A p88

PUPIL TEXTBOOK 4A PAGE 123

Practice

WAYS OF WORKING Independent thinking

IN FOCUS In question ③, children calculate the exact answer and then use an estimate to check it. This will help them with their reasoning.

STRENGTHEN The matching exercise in question ② will provide support for children who need it, and will allow them to focus on developing their reasoning skills.

DEEPEN Focus on question ④ and encourage discussion about rounding to different degrees of accuracy. Ask: *Can you think of real-life situations when the degree of accuracy is important?* For example, estimating the rough total price of a shopping list.

ASSESSMENT CHECKPOINT Question ② will allow you to assess which children can choose a suitable estimate. They should also be able to explain why you might sometimes round a number in the 1,000s to the nearest 100.

ANSWERS Answers for the **Practice** part of the lesson can be found in the *Power Maths* online subscription.

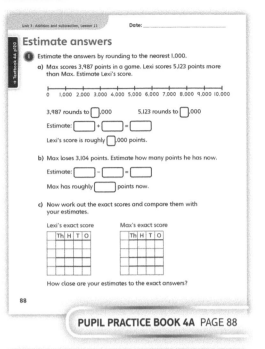

PUPIL PRACTICE BOOK 4A PAGE 88

PUPIL PRACTICE BOOK 4A PAGE 89

PUPIL PRACTICE BOOK 4A PAGE 90

Reflect

WAYS OF WORKING Pair work

IN FOCUS For this question, some children may calculate 2,000 – 1,000, whilst others may calculate 1,915 – 1,000 and some others may calculate 2,000 – 1,019. Encourage children to share their solutions and debate which is the most useful.

ASSESSMENT CHECKPOINT This will help you to assess children's understanding of the methodology used in this lesson. Can they list the instructions in clear steps?

ANSWERS Answers for the **Reflect** part of the lesson can be found in the *Power Maths* online subscription.

After the lesson

- Can all children estimate using rounding?
- How many children achieved mastery of this lesson?
- Do children understand why we estimate answers?

Check strategies

Learning focus

In this lesson, children will learn strategies for checking answers, using the inverse operation and estimating by rounding.

Before you teach

- Do all children know what 'inverse' means?
- How will you link this lesson to the previous one?
- Do children understand the importance of checking an answer?

NATIONAL CURRICULUM LINKS

Year 4 Number – addition and subtraction

Estimate and use inverse operations to check answers to a calculation.

ASSESSING MASTERY

Children can complete a calculation and then use the inverse operation to check their answer. They can spot mistakes and understand the importance of checking answers, and can understand that there is more than one way to check an answer (inverse, rounding, repetition).

COMMON MISCONCEPTIONS

Children may work out the inverse, but if it is not the same as their answer, they may not know what to do. Ask:
- *Could you work out the calculation in a different way?*

STRENGTHENING UNDERSTANDING

Some children may not be secure with inverse operations. Scaffold learning by showing simple fact families and explaining the relationship between them, for example $3 + 2 = 5$, $2 + 3 = 5$, $5 - 2 = 3$, $5 - 3 = 2$.

GOING DEEPER

Deepen learning in this lesson by exploring different ways to check answers. Children may find the inverse, or they may use rounding or repetition. Ask: *Can you explain the different strategies? What are the differences between the strategies?*

KEY LANGUAGE

In lesson: check, addition, subtraction, inverse, fact family

Other language to be used by the teacher: round, nearest, thousands (1,000s), hundreds (100s), tens (10s), ones (1s)

STRUCTURES AND REPRESENTATIONS

Bar model, part-whole model

 In the eTextbook of this lesson, you will find interactive links to a selection of teaching tools.

Quick recap

Write five different calculations on the board. Three of them should be correct and two should contain an error.

Examples of possible errors include:

$2,500 - 1,501 = 4,001$ (wrong operation)

$1,350 + 750 = 2,000$ (exchanged incorrectly)

Challenge children to find and explain the correct answers.

Discover

WAYS OF WORKING Pair work

ASK

- Question ① a): *How many different ways can you think of to check an answer?*
- Question ① a): *What would you do if you got a different answer when checking?*

IN FOCUS Question ① a) focuses on the importance of checking answers. If the checked answer is different, we know there is a problem but is it the original or the checked answer that is incorrect? Discuss this with children.

PRACTICAL TIPS Create a display in the classroom that models different ways to check answers.

ANSWERS

Question ① a): A subtraction can be checked by using the inverse operation, which is addition.
799 + 574 = 1,373
The parts do not match the whole.
The calculation should be done again.

Question ① b): 1,225 – 799 = 426; 1,226 – 800 = 426
There are 426 km left to travel.

Check strategies

Discover

We had 1,225 km to travel. We have travelled 799 km already. I calculate we have 574 km left.

I will check.

① a) How can the astronaut check her calculation?

b) Show two ways to do the calculation.

124

PUPIL TEXTBOOK 4A PAGE 124

Share

WAYS OF WORKING Whole class teacher led

ASK

- Question ① b): *Could you use equivalent difference to work out the answer?*
- Question ① b): *Why is it important to check answers?*

IN FOCUS For question ① b), it is a good idea to remind children of the earlier lesson in this unit about equivalent difference. Explain that there are different ways to check answers, and sometimes you may even need to use more than one to be sure. Link this to real-life examples, for example a shopkeeper totalling their takings for a day.

Share

a) A subtraction can be checked by using the inverse operation, which is addition.

1,225	
799	574

	Th	H	T	O
		7	9	9
+		5	7	4
	1	3	7	3
		1	1	

I used the fact family to check by adding the parts.

I checked by estimating.

The parts do not match the whole. The astronaut's calculation should be done again.

b)

Th	H	T	O
X̶	¹2̶	¹2̶	¹5
–	7	9	9
	4	2	6

Th	H	T	O
X̶	¹2	2	6
–	8	0	0
	4	2	6

I found an easier way. I added 1 to each number.

There are 426 km left to travel.

125

PUPIL TEXTBOOK 4A PAGE 125

Think together

Unit 3: Addition and subtraction, Lesson 12

Think together

WAYS OF WORKING Whole class teacher led (I do, We do, You do)

ASK

• Question **2**: *What sort of calculation could you do to check the answers accurately? How could you estimate first what the answers should be?*

IN FOCUS Question **1** requires children to demonstrate that they can identify the inverse operation to use to check a calculation in the context of a word problem. In this instance they will need to do an addition to check, and they should find that the original calculation is correct.

STRENGTHEN Following on from question **2**, give children more opportunities to check answers by using the inverse operation.

DEEPEN Ask children to complete the calculations in question **2**, then challenge them to explain where each one went wrong. For the first one, children should be able to do 5,391 – 3,401 = 1,990 to show that a 0 was missing from the original calculation. Children should find that the second one is already correct.

ASSESSMENT CHECKPOINT Question **1** gives a simple way to assess children on whether they can check answers using the inverse operation. Question **3** will allow you to assess which children can check answers using visual representations such as bar models or part-whole models.

ANSWERS

Question **1**: 3,288 + **3,707** = 6,995
The parts **do** match the whole. The calculation is correct.

Question **2** a): 5,391 – 3,401 = 1,990 (Correction: either
199 + 3,401 = 3,600
or 1,990 + 3,401 = 5,391)

Question **2** b): 8,569 + 440 = 9,009

Question **3**: Look for accurately drawn part-whole models or bar models showing:
1,090 + 1,910 = 3,000   4,000 – 2,750 = 1,250
2,550 = 700 + 1,850   2,750 – 750 = 2,000

Think together

1 The mass of the astronaut's food has to be calculated accurately. Check the calculation using the inverse operation.

> 6,995 g of food at start of voyage.
> 3,288 g eaten so far.
> 6,995 – 3,288 = 3,707

6,995 / 3,288 / ◯

The parts **do / do not** match the whole.
The calculation **is / is not** correct.

2 Write a calculation to check each of these.

a)
199 + 3,401 = 5,391

b)
9,009 – 440 = 8,569

Complete any corrections that are needed.

126

PUPIL TEXTBOOK 4A PAGE 126

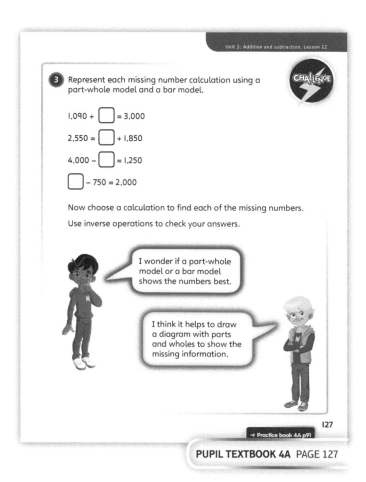

3 Represent each missing number calculation using a part-whole model and a bar model.

1,090 + ◻ = 3,000

2,550 = ◻ + 1,850

4,000 – ◻ = 1,250

◻ – 750 = 2,000

Now choose a calculation to find each of the missing numbers.
Use inverse operations to check your answers.

I wonder if a part-whole model or a bar model shows the numbers best.

I think it helps to draw a diagram with parts and wholes to show the missing information.

127

→ Practice book 4A p91

PUPIL TEXTBOOK 4A PAGE 127

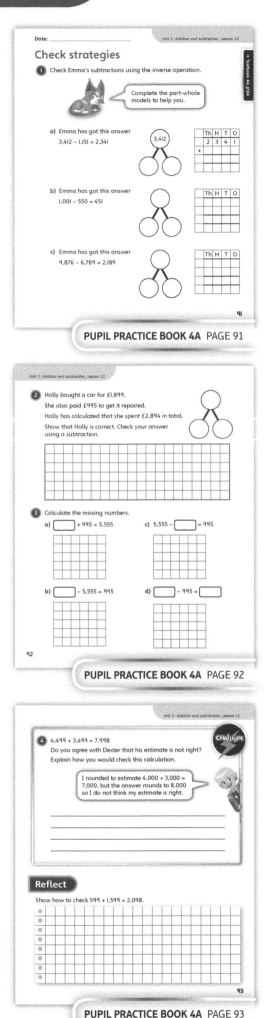

Practice

WAYS OF WORKING Independent thinking

IN FOCUS Checking answers, such as in question **1**, can empower and motivate children as they enjoy becoming like the teacher.

STRENGTHEN Question **3** features missing numbers in calculations. Some children's knowledge may need strengthening here. Ask: *When do you need to use the inverse operation to find the answer? When don't you need to use the inverse operation? What other methods for checking can you use?*

DEEPEN Deepen learning in question **1** by asking: *What mistakes did Emma make? Can you explain why?*

THINK DIFFERENTLY Question **4** shows how rounding can sometimes be flawed. First, encourage children to look at the correct answer (7,998). Then see if they can think why the rounding method used was not accurate. Children should reason that Dexter would have been more accurate if he had rounded to the nearest hundred instead of thousand.

ASSESSMENT CHECKPOINT Question **2** will allow you to assess whether children can choose an appropriate method for checking the answer to a calculation.

ANSWERS Answers for the **Practice** part of the lesson can be found in the *Power Maths* online subscription.

Reflect

WAYS OF WORKING Pair work

IN FOCUS This is a good opportunity for discussion using maths language. Ask children to work with a partner for this activity, and to work together to discuss their methods for checking.

ASSESSMENT CHECKPOINT This activity will let you see which children can use more than one strategy to check an answer.

ANSWERS Answers for the **Reflect** part of the lesson can be found in the *Power Maths* online subscription.

After the lesson ⏸

- Can children use more than one method to check answers?
- Can children reason why it is important to check answers?
- Are children ready to apply their knowledge to lessons about problem solving?

PUPIL PRACTICE BOOK 4A PAGE 91

PUPIL PRACTICE BOOK 4A PAGE 92

PUPIL PRACTICE BOOK 4A PAGE 93

Problem solving – one step

Learning focus

In this lesson, children will apply addition and subtraction strategies they have learnt previously to solve simple problems.

Before you teach

- Are children ready to move on to problem solving?
- How will you draw out the key vocabulary in the lesson?
- How will you promote discussion of methods in this lesson?

NATIONAL CURRICULUM LINKS

Year 4 Number – addition and subtraction

Solve addition and subtraction two-step problems in contexts, deciding which operations and methods to use and why.

ASSESSING MASTERY

Children can choose an efficient method of addition or subtraction to solve a problem. They can represent the problem on a bar model, and explain their method.

COMMON MISCONCEPTIONS

Children may struggle to interpret the word problems and not know whether to add or subtract. Ask:
- *Can you highlight the key words that might help you?*

STRENGTHENING UNDERSTANDING

Some children may need practice in interpreting word problems. Talk through the questions and provide bar models to visually represent them.

GOING DEEPER

Provide more examples of problems where numbers are represented by symbols (early algebra). Discuss the best way for children to show what information they have and what they need to find out.

KEY LANGUAGE

In lesson: problem solving, strategy, part, whole, bar model, diagram, story problem, altogether, left

STRUCTURES AND REPRESENTATIONS

Part-whole model, bar model

RESOURCES

Strips of paper to make bar models

 In the eTextbook of this lesson, you will find interactive links to a selection of teaching tools.

Quick recap

Ask children to invent their own word problems for the following calculations:

25 + 50   150 – 95

Discover

ASK

- Question ① a): *Which model will you use?*
- Question ① a): *Would a bar model represent the problem well?*

IN FOCUS In question ① a), children have to represent a word problem visually. If children are not sure which diagram to choose, suggest the bar model or part-whole model.

PRACTICAL TIPS Instead of drawing bar models, children could cut strips of paper to make physical bar models.

ANSWERS

Question ① a):   Total votes

2,899   ?

No votes   Yes votes

Total votes
5,762

2,899	?

No votes   Yes votes

Question ① b): You need to subtract to find the missing part and calculate the answer:
5,762 − 2,899 = 2,863.
There were 2,863 Yes votes. No got more votes because 2,899 > 2,863.

Share

ASK

- Question ① a): *What does 'part' mean?*
- Question ① a): *What does 'whole' mean?*

IN FOCUS For question ① a), the bar model and the part-whole model have been used to represent the problem. Be aware that children may have used other representations, for example number lines.

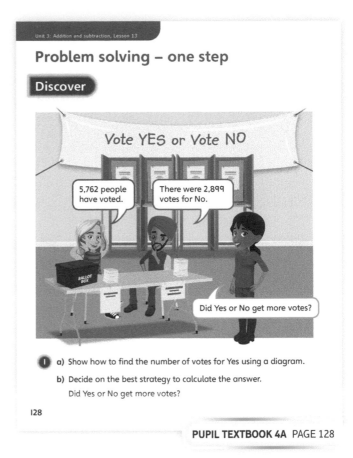

PUPIL TEXTBOOK 4A PAGE 128

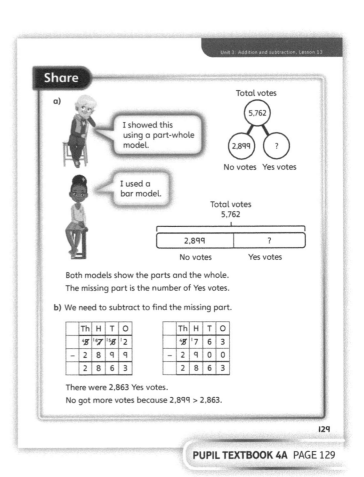

PUPIL TEXTBOOK 4A PAGE 129

Think together

Whole class teacher led (I do, We do, You do)

ASK

- Question **1**: *How do you know that this is an addition problem?*
- Question **2**: *What does the bar model need to show?*
- Question **4**: *What do you need to remember when drawing a bar model?*

IN FOCUS Question **3** will draw out associated addition and subtraction facts. Explain to children that making connections like this is very important in maths mastery.

STRENGTHEN After completing question **2**, children could strengthen their learning by drawing the correct bar model once they have identified what is wrong with the two that are given.

DEEPEN Use question **4** to deepen learning. Children will need to represent a range of calculations using bar models, including some with missing numbers. Challenge children to find links between the calculations. They should be able to use vocabulary such as 'inverse', and explain the strategies they used to work out the correct answers. Finally, they should realise that they do not need four different bar models since the calculations are linked, and so only two models are needed. Extend learning by providing an answer, such as 3,232, and asking children to draw bar models to match it.

ASSESSMENT CHECKPOINT Question **2** will allow you to assess which children can accurately represent a word problem with a bar model. Look for effective reasoning and the correct use of mathematical vocabulary.

ANSWERS

Question **1**: 1,775 (Yes); 3,007 (No).
3,007 + 1,775 = 4,782. 4,782 people voted.

Question **2**: 2,111 people are too young to vote.
Jamilla has put 9,923 as a part when it should be the whole.
Max has drawn the correct bar model, but the parts should not be equally sized.

Question **3**: 6,000 − 2,999 = 3,001
6,000 − 3,001 = 2,999
2,999 + 3,001 = 6,000
3,001 + 2,999 = 6,000

Question **4**: Look for accurately drawn bar models, with parts and whole of appropriate sizes. Children should notice that only two models are needed, showing:
2,674 − 199 = 2,475 and 199 + 2,475 = 2,674
2,475 − 199 = 2,276 and 199 = 2,475 − 2,276

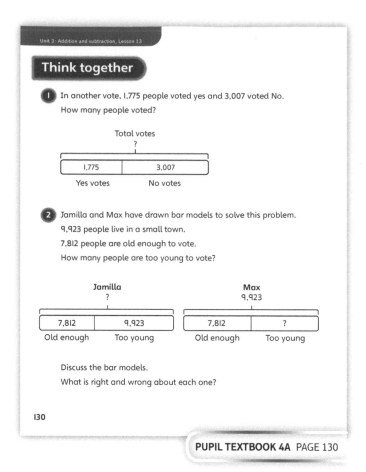

PUPIL TEXTBOOK 4A PAGE 130

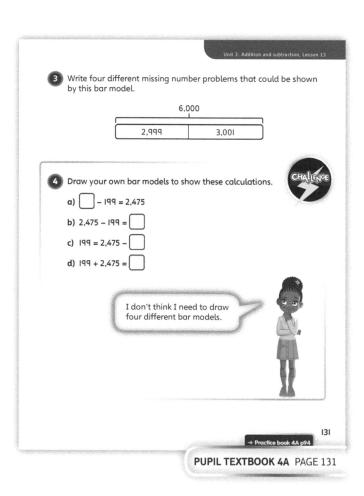

PUPIL TEXTBOOK 4A PAGE 131

Practice

WAYS OF WORKING Independent thinking

IN FOCUS Question ② promotes reasoning. Ask children how they will go about solving each problem. Also ask them what their bar model will tell them, and what the size of each part should be.

STRENGTHEN For question ④, talk to children about how they could begin to find a solution. They may realise that starting with the triangle would be a good strategy. From the second bar model: if you add the triangle to the star you get the cloud. But also, we know that if you add 2,000 to the star you get the cloud. So the triangle must be worth 2,000. From the first model, that means that the star and the heart sum to 2,000. As the heart is worth 1,000 more than the star, the heart must be 1,500 and the star 500. This means that the cloud is 500 + 2,000 = 2,500.

DEEPEN In question ④, tell children that you think two hearts add up to less than one triangle. Challenge children to reason why this is incorrect (heart > star; star + heart = triangle).

ASSESSMENT CHECKPOINT Assess children's progress by looking at question ②. See if they have achieved mastery, and can interpret the question correctly and use the appropriate operation.

ANSWERS Answers for the **Practice** part of the lesson can be found in the *Power Maths* online subscription.

Reflect

WAYS OF WORKING Pair work

IN FOCUS Some children may need support with thinking of a suitable context for their problem. If needed, provide some ideas like measuring in centimetres, or counting marbles in a jar.

ASSESSMENT CHECKPOINT This activity will let you see which children are likely to have mastered the lesson. They will be able to create a relevant story problem and explain how it relates to the bar model.

ANSWERS Answers for the **Reflect** part of the lesson can be found in the *Power Maths* online subscription.

After the lesson ⏸

- Can all children represent addition and subtraction problems using a bar model?
- Can they explain the bar model using the words 'part' and 'whole'?
- Are children ready to move on to more complex problem solving?

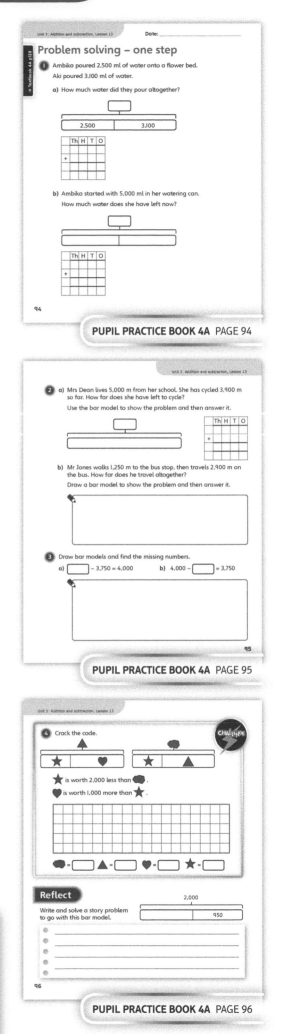

PUPIL PRACTICE BOOK 4A PAGE 94

PUPIL PRACTICE BOOK 4A PAGE 95

PUPIL PRACTICE BOOK 4A PAGE 96

Problem solving – comparison

Learning focus

In this lesson, children will explore single bar models and comparison bar models to interpret and solve simple problems.

Before you teach

- How will you explain the difference between a single bar model and a comparison bar model?
- Will you have a challenge activity for any quick finishers?
- Could you make a classroom display to support this lesson?

NATIONAL CURRICULUM LINKS

Year 4 Number – addition and subtraction

Solve addition and subtraction two-step problems in contexts, deciding which operations and methods to use and why.

ASSESSING MASTERY

Children can understand when to draw single bar models and when to draw comparison bar models to help them solve problems. They can explain the representations clearly using the correct vocabulary.

COMMON MISCONCEPTIONS

Children may not understand when to draw single bar models and when to draw comparison bar models. Ask:
- *Can you draw both models and explain which is more useful?*

STRENGTHENING UNDERSTANDING

If children do not know when to draw a single bar model and when to draw a comparison bar model, give them some word problems with corresponding single bar models and comparison bar models. Ask them to discuss as a group which one represents the problem more clearly.

GOING DEEPER

Deepen learning by giving children some single bar models and comparison bar models, and ask them to write some word problems to match them.

KEY LANGUAGE

In lesson: problem solving, addition, subtraction, single bar model, comparison bar model, part, whole, story problem, how much, more, fewer, left, difference, more than

Other language to be used by the teacher: strategy

STRUCTURES AND REPRESENTATIONS

Single bar model, comparison bar model

 In the eTextbook of this lesson, you will find interactive links to a selection of teaching tools.

Quick recap 𝒬

Ask children to write out this list of 3-digit numbers in order from smallest to greatest:

501, 244, 910, 550, 243, 509

Discover

Pair work

ASK

- Question ❶ a): *How will you show that Luis has more?*
- Question ❶ b): *Will you use column subtraction or is there another way of calculating the answer?*

IN FOCUS Questions ❶ a) and ❶ b) explore the difference between single bar models and comparison bar models. If necessary, suggest that the bar models may look different for each question.

PRACTICAL TIPS Give children visual support, by creating a classroom display featuring some word problems with single bar models and comparison bar models, to support their interpretation of them.

ANSWERS

Question ❶ a): A single bar model or comparison bar model where the part representing 1,005 (Luis) is noticeably longer than the part representing 899 (Danny).

Look for children who identify that the comparison bar model is a better way to represent this problem.

Question ❶ b): 1,005 – 899 = 106. Luis has 106 more points than Danny.

PUPIL TEXTBOOK 4A PAGE 132

Share

Whole class teacher led

ASK

- Question ❶ a): *How does the comparison bar model show who has more?*
- Question ❶ b): *Did you use the column method to calculate the answer, or a different strategy?*

IN FOCUS For question ❶ b), children may have used a different strategy, for example, equivalent difference. Share all the strategies that children used.

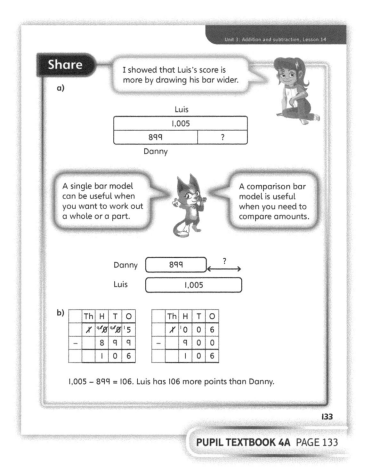

PUPIL TEXTBOOK 4A PAGE 133

Think together

WAYS OF WORKING Whole class teacher led (I do, We do, You do)

ASK

- Question **2**: *Is this problem an addition or a subtraction?*
- Question **2**: *What can you fill in first on the model?*
- Question **3**: *Can you think of more than one solution for part d)?*

IN FOCUS Question **2** breaks the word problem down for children. Ask them to fill in the information that they know first, then to explain what calculation they must do to find the answer.

STRENGTHEN Strengthen learning by talking through question **3**. Help children to identify that a comparison bar model is needed when finding a difference, and a single bar model is more useful when completing an addition.

DEEPEN Ask children to actually draw the different bar models for question **3** and to reason which is more suitable for each question. For the final question, challenge children to find more than one answer.

ASSESSMENT CHECKPOINT Question **3** will let you see which children have mastered the lesson and can identify the correct bar model and calculate the answers effectively.

ANSWERS

Question **1**: 1,050 – 678 = 372. Amelia has 372 more points than Jack.

Question **2**: 975 + 875 = 1,850. Isla has 1,850 points.

Question **3** a): Single bar model showing
5,250 + 100 = 5,350

Question **3** b): Single bar model showing
5,250 + 750 = 6,000

Question **3** c): Comparison bar model showing
5,250 – 750 = 4,500

Question **3** d): Comparison bar model showing any two numbers with a difference of 2,000.

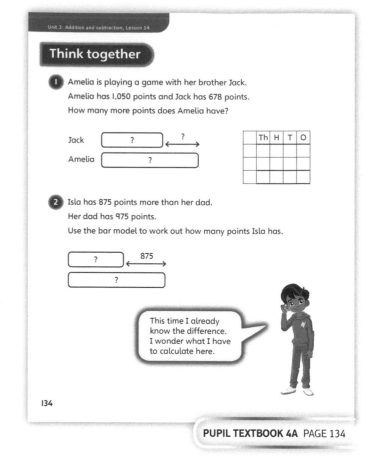

PUPIL TEXTBOOK 4A PAGE 134

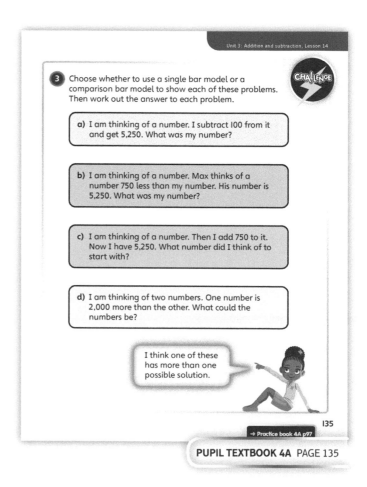

PUPIL TEXTBOOK 4A PAGE 135

Practice

WAYS OF WORKING Independent thinking

IN FOCUS Questions ❶ a), ❶ b) and ❶ c) draw out the different ways that additions and subtractions can be represented using bar models.

STRENGTHEN Some children will need scaffolding for question ❹. Work alongside them, or as a group, and talk through the problem. You could provide some of the bars for them.

DEEPEN In question ❸, challenge children to find more than one way to work out the correct answer.

THINK DIFFERENTLY Question ❷ may prove challenging for some children. Usually the word 'more' is associated with addition. However, in this question you have to find how many 'more than'. Explain the difference to children to help them identify that this is a subtraction. Use a bar model to support this.

ASSESSMENT CHECKPOINT Question ❹ will tell you which children have mastered the lesson. Check that they can break down the problem and represent it using bar models. Ask them to explain their workings to make sure they are confident with their learning.

ANSWERS Answers for the **Practice** part of the lesson can be found in the *Power Maths* online subscription.

Reflect

WAYS OF WORKING Pair work

IN FOCUS Display the words 'compare', 'comparison', 'addition', 'subtraction' and 'difference' to help children explain their answers. Encourage children to use examples of number sentences and bar models in their explanations.

ASSESSMENT CHECKPOINT Assess children on the accuracy of their explanations, and the vocabulary they use.

ANSWERS Answers for the **Reflect** part of the lesson can be found in the *Power Maths* online subscription.

After the lesson ⏸

- Do all children know the difference between a single bar model and a comparison bar model?
- Which children struggled with **Practice** question ❹?
- Could you do some more problems like this to strengthen learning?

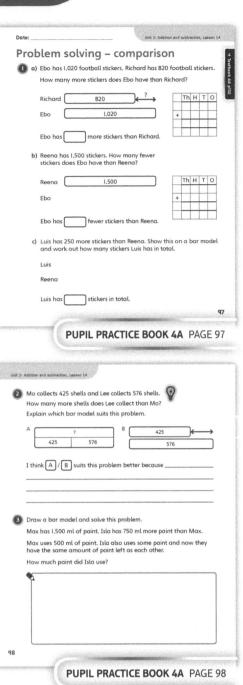

PUPIL PRACTICE BOOK 4A PAGE 97

PUPIL PRACTICE BOOK 4A PAGE 98

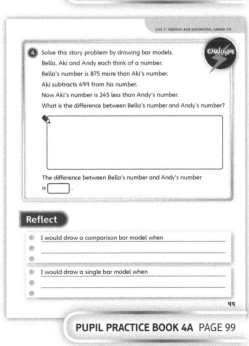

PUPIL PRACTICE BOOK 4A PAGE 99

Problem solving – two steps

Learning focus

In this lesson, children will apply addition and subtraction strategies that they have learnt previously, to solve two-step problems.

Before you teach

- How will you scaffold learning in this lesson?
- Could you pair children who struggle with reading with a more capable reader?
- How can you promote maths vocabulary in this lesson?

NATIONAL CURRICULUM LINKS

Year 4 Number – addition and subtraction

Solve addition and subtraction two-step problems in contexts, deciding which operations and methods to use and why.

ASSESSING MASTERY

Children can understand how to solve a problem, and which operations they must use. They can confidently represent a word problem using a model or representation and explain this clearly.

COMMON MISCONCEPTIONS

Children may not realise that a question is a multi-step problem, and so may only complete one of the steps. Ask:
- *Is that your final answer?*

STRENGTHENING UNDERSTANDING

Encourage children to break each problem down into simple steps. Each step should be supported with a model or representation, helping children to interpret the mathematics.

GOING DEEPER

Give children two completed bar models and challenge them to think of a multi-step word problem to match.

KEY LANGUAGE

In lesson: problem solving, addition, subtraction, step, check, part, whole, bar model, story problem, total, difference, how much

Other language to be used by the teacher: multi-step, strategy

STRUCTURES AND REPRESENTATIONS

Bar model

 In the eTextbook of this lesson, you will find interactive links to a selection of teaching tools.

Quick recap

Ask children to find the answer to this calculation:

9 + 19 + 29

Discuss and compare the methods that they used.

Discover

Pair work

ASK

- Question **1** a): *Is there only one way to work this out?*
- Question **1** a): *What number sentences will help you work out the correct answer?*

IN FOCUS Question **1** a) involves children working out a multi-step problem. Focus children's learning by asking them to break the question down into two steps. They should write two number sentences and then do the calculations. There is more than one way to work out the answer. Encourage children to explore all of the options.

PRACTICAL TIPS Promote lots of discussion in this section, using appropriate maths vocabulary. Talking through questions in pairs can help to ensure that children gain a secure understanding of the problem.

ANSWERS

Question **1** a): 2,500 – 1,200 – 750 = 550
or 1,200 + 750 = 1,950 and 2,500 – 1,950 = 550
Olivia will need to run 550 m to complete the race.

Question **1** b): Check by adding.
1,200 + 750 = 1,950 and 1,950 + 550 = 2,500
or 1,200 + 750 + 550 = 2,500

PUPIL TEXTBOOK 4A PAGE 136

Share

Whole class teacher led

ASK

- Question **1** b): *How can you check your answer?*
- Question **1** b): *Can you think of more than one way to check?*

IN FOCUS For question **1** b), children should check by using the inverse operation (addition). However, some children may choose to check by using a different strategy, for example, a number line or rounding.

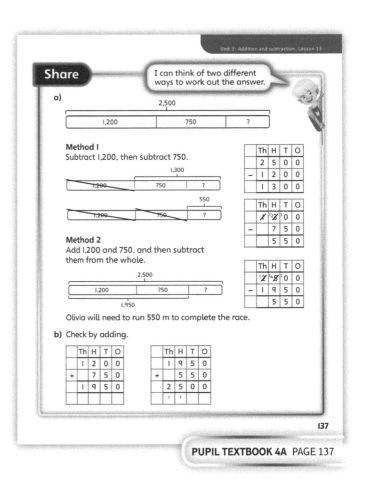

PUPIL TEXTBOOK 4A PAGE 137

Think together

ASK

- Question ➊ a): *Have you seen a similar question before?*
- Question ➊ b): *What is different about the bar model here?*
- Question ➌: *How can you show two numbers with a total of 1,500 but a difference of 1,000?*

IN FOCUS In question ➊ b), children will see that the whole is at the side of the bars, which is new. Explain that this is just a different way of looking at a bar model. Ask children to draw what this would usually look like, i.e. a single bar model with the whole as 2,475 and three parts of 475, 800 and 1,200.

STRENGTHEN Question ➊ a) is similar to the **Discover** section. This will support children, so prompt them to think back to this if necessary. Ask children to find both ways of working out the problem.

DEEPEN Question ➊ b) is a good chance to show children that you do not always have to use a written method when calculating. Some children will realise that a good strategy would be to add 1,200 and 800 to reach 2,000, then add on the 475.

ASSESSMENT CHECKPOINT Question ➋ will allow you to assess which children can use a comparison bar model to represent a multi-step problem. Children may break it down into a number of steps, first adding 1,250 and 300 to get 1,550, then adding 1,250 and 1,550 to get 2,800, and finally subtracting 2,800 from 3,000 to get 200.

ANSWERS

Question ➊ a): 5,000 – 1,250 – 1,750 = 2,000.
Toshi ran 2,000 m.

Question ➊ b): 475 + 800 + 1,200 = 2,475.
Toshi, Amal and Jen ran 2,475 m in total.

Question ➋: 1,250 + 300 = 1,550 watched the javelin.
3,000 – 1,550 = 1,450 watched the long jump.

Question ➌: Look for comparison bar models showing the first bar as 1,500 and the second bar as 1,000 plus two equal blank boxes; then a deduction that the blanks are half of 500, or 250.
So Emma ran 1,250 m and Alex ran 250 m.

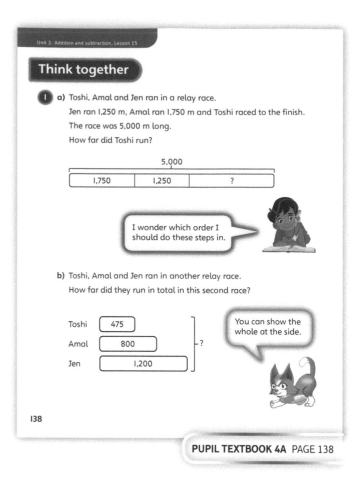

PUPIL TEXTBOOK 4A PAGE 138

PUPIL TEXTBOOK 4A PAGE 139

Practice

WAYS OF WORKING Independent thinking

IN FOCUS Question **4** a) is tricky because we are not told how much money Amy or Ben actually have. Scaffold learning by asking children to draw a bar model, or to write down the number sentences. You could start to introduce some simple algebra by representing Amy and Ben with the letters A and B. For example:

A has £1,275 less than B.

B spends £550.

£1,275 – £550 = £725

A now has £725 less than B.

A gets £750.

£750 – £725 = £25

A now has £25 more than B.

STRENGTHEN For question **1** b), it may be necessary to explain what a triathlon is (a race with three parts: a swim, a cycle and a run). Question **4** b) can be linked to knowledge of number bonds. Support learning by asking: *What is left if you take 800 from 2,800? Could you partition this number to help you?* They could represent this with bar models.

DEEPEN Question **2** displays the use of the bar model in a real-life context. Discuss this and see if children can think of other similar applications, for instance, perimeter.

ASSESSMENT CHECKPOINT Question **4** will allow you to assess which children have a firm understanding of interpreting multi-step word problems. Look in particular at how they have shown their workings.

ANSWERS Answers for the **Practice** part of the lesson can be found in the *Power Maths* online subscription.

Reflect

WAYS OF WORKING Pair work

IN FOCUS Challenge children to think of as many solutions as they can. You could extend learning by asking them to think of a subtraction with three numbers that equal 2,050, or even an addition and subtraction with three numbers, such as 1,000 + 2,000 – 950.

ASSESSMENT CHECKPOINT Assess children on using the correct sizes for the bars in their models and for demonstrating depth of thinking, for example not choosing a very simple example like 2,048 + 1 + 1.

ANSWERS Answers for the **Reflect** part of the lesson can be found in the *Power Maths* online subscription.

After the lesson

- How did children approach **Practice** question **4**?
- Do some children need more support with visualising problems like this?
- Are children ready to move on to the final lesson of the unit?

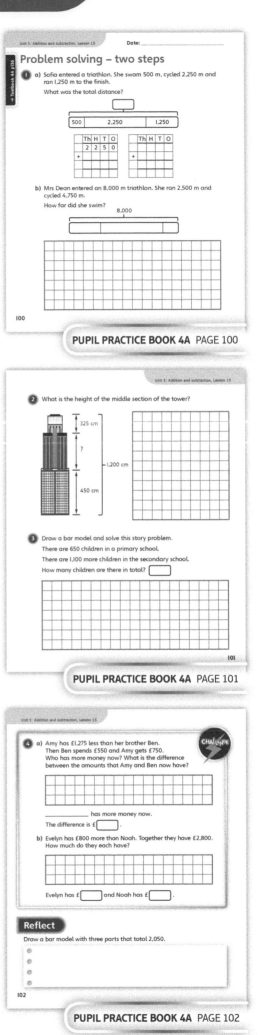

PUPIL PRACTICE BOOK 4A PAGE 100

PUPIL PRACTICE BOOK 4A PAGE 101

PUPIL PRACTICE BOOK 4A PAGE 102

Problem solving – multi-step problems

Learning focus

In this lesson, children will continue to apply the addition and subtraction strategies that they have previously learnt to solve multi-step problems.

Before you teach

- Will you provide any practical resources in this lesson?
- How did the previous lesson go?
- Are children ready for more complex problems?

NATIONAL CURRICULUM LINKS

Year 4 Number – addition and subtraction

Solve addition and subtraction two-step problems in contexts, deciding which operations and methods to use and why.

ASSESSING MASTERY

Children can understand how to solve a problem, and can identify which operations they must use. They can confidently represent a multi-step problem using models or representations and can explain them clearly.

COMMON MISCONCEPTIONS

Children may not correctly interpret the questions. Ask:
- *Did you use the correct operations? How can you check?*

STRENGTHENING UNDERSTANDING

To strengthen learning in this lesson, run a quick intervention in which you talk through and explain some word problems. Support the discussion with representations to make the problems visual.

GOING DEEPER

Deepen learning by providing some more bar models with numbers but no labels, and asking children to write a story problem to go with them. Encourage children to use real-life contexts.

KEY LANGUAGE

In lesson: problem solving, addition, subtraction, diagram, bar model, greater than, less than, step, total, part, whole, story problem, how much, left

Other language to be used by the teacher: multi-step, strategy

STRUCTURES AND REPRESENTATIONS

Part-whole model, bar model, number line

 In the eTextbook of this lesson, you will find interactive links to a selection of teaching tools.

Quick recap

Ask children to complete this calculation:

9 + 99 + 999 + 1,999

Discuss and compare the methods that they used.

Discover

Pair work

ASK

- Question ❶ a): *Can you spot any information that is not needed?*
- Question ❶ a): *Is there more than one way to work this out?*

IN FOCUS In this question, some 'useless' information is also provided. Ask children if they need to know about Camp 4. They should be able to reason that the information about Camp 4 does not help to answer the question.

PRACTICAL TIPS Encourage children to draw the problem. Having an image of the mountain with the bases and the mountaineer in front of them will allow children to interpret the question more easily.

ANSWERS

Question ❶ a): Children may draw a bar model, a number line or a part-whole model to show this. Look for diagrams that show a total of 5,275 and three parts of 2,450, 1,500 and ☐.

Question ❶ b): To solve this, find out if the distance from Jen to Camp 2 is greater than or less than 1,500 m.
2,450 + 1,500 = 3,950
5,275 − 3,950 = 1,325
1,500 > 1,325, so Jen is closer to Camp 2.

Problem solving – multi-step problems

Discover

Base camp 0 m
Camp 1   2,450 m
Camp 2   5,275 m
Camp 3   7,299 m
Camp 4   8,158 m

Jen   Toshi   Amal

❶ a) Jen has climbed 1,500 m higher than Camp 1.
   She wants to know if she is now closer to Camp 1 or Camp 2.
   How could she show the answer with a diagram?

b) Is Jen closer to Camp 1 or Camp 2?

140

PUPIL TEXTBOOK 4A PAGE 140

Share

Whole class teacher led

ASK

- Question ❶ b): *What does 'closer' mean?*
- Question ❶ b): *How could you use the signs < or > in your explanation?*

IN FOCUS Children may mistakenly think that Jen is closer to Camp 1 because 1,500 is greater than 1,325 and children often assume that the correct answer is the one that is higher. Explain what 'closer' means and then ask which is closer, something 1,500 m away or something 1,325 m away.

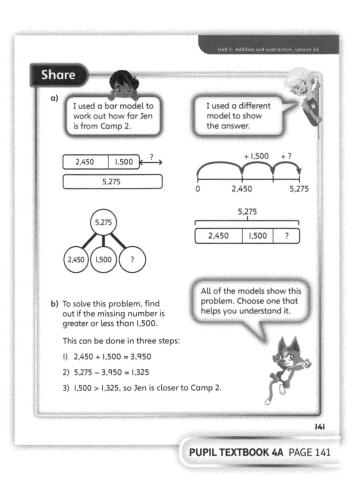

Share

a)

I used a bar model to work out how far Jen is from Camp 2.

I used a different model to show the answer.

| 2,450 | 1,500 | ? →
| 5,275 |

+ 1,500   + ?

0   2,450   5,275

5,275

| 2,450 | 1,500 | ? |

5,275

(2,450) (1,500) (?)

b) To solve this problem, find out if the missing number is greater or less than 1,500.

This can be done in three steps:

1) 2,450 + 1,500 = 3,950
2) 5,275 − 3,950 = 1,325
3) 1,500 > 1,325, so Jen is closer to Camp 2.

All of the models show this problem. Choose one that helps you understand it.

141

PUPIL TEXTBOOK 4A PAGE 141

Think together

Whole class teacher led (I do, We do, You do)

ASK

- Question **1**: *Why do the other diagrams not match the problem?*
- Question **3**: *How can you solve this problem in steps?*

IN FOCUS You may need to break question **3** down to scaffold learning. Ask children to draw the mountains with the differences in heights marked to help them gain a visual understanding of the problem. Then ask children to convert this information into bar models, as it is more mathematical to work like this. Then ask them to calculate and mark the totals.

STRENGTHEN Strengthen learning in question **1** by asking children to discuss what each diagram shows, and then reason which is the correct one. Ask children why the others do not match the question. You could ask them to think of word problems that would match each diagram.

DEEPEN Ask children to make up their own multi-step problems based on the number sentence 453 + 234 – 101.

ASSESSMENT CHECKPOINT Question **3** will allow you to check which children can interpret a problem, break it down into steps, represent it with diagrams and solve it.

ANSWERS

Question **1**: Bar model A shows the problem.
$$1,245 - 385 = 860$$
$$1,245 + 860 = 2,105$$
Amal has climbed 2,105 m in total.

Question **2**: Look for a diagram that shows:
$$6,895 - 1,812 - 1,259 - 2,248 = 1,576$$
Jen climbed 1,576 m on Day 3.

Question **3**: $3,466 + 1,344 = 4,810$
Mont Blanc is 4,810 m high.
$$4,810 - 1,030 = 3,780$$
Mount Fuji is 3,780 m high.

Think together

1 Amal climbed 1,245 m on Day 1, and then climbed 385 m less than that on Day 2.

He wants to work out how far he has climbed in total.

Decide which model shows this problem, and then solve it.

A
| 1,245 |
| ? | 385 → | } ?

C
+1,245   +385
0   ?

B
1,245
385   ?

D
?
| 1,245 | 385 | ? |

2 The mountain is 6,895 m high.

After reaching the top, Jen climbed back down.

She climbed down 1,812 m on Day 1 and 1,259 m on Day 2.

After Day 3, she had 2,248 m left to climb.

Draw a diagram to show how far Jen climbed on Day 3, then calculate the answer.

> I will try to work out if this is about parts and wholes, or comparing amounts.

142

PUPIL TEXTBOOK 4A PAGE 142

3

Ben Nevis

1,344 m

Ben Nevis is the tallest mountain in the UK.

Ben Nevis is 3,466 m shorter than Mont Blanc and Mont Blanc is 1,030 m taller than Mount Fuji.

Calculate the height of Mount Fuji.

Use a model to show the problem.

> This problem has two steps, so I will draw two different models.

> We are comparing three mountains. I wonder if I can draw three bars in one model.

143

→ Practice book 4A p103

PUPIL TEXTBOOK 4A PAGE 143

Practice

WAYS OF WORKING Independent thinking

IN FOCUS In question **2**, some children may just do 3,985 – 1,700 and give 2,285 as their final answer. Discuss what 'fewer' means and also what 'total' means. This should help children to realise that they then have to add 3,985 and 2,285.

STRENGTHEN In question **1**, children need to add three amounts. Strengthen learning by modelling how they can do this using the column method, as they may not have come across this before. Then set other additions to practise this.

DEEPEN In question **4**, children are required to interpret a diagram and think of a matching story problem. Ask them to explain the diagram to you and to fill in the missing amounts for Class 1 and Class 2. Deepen learning by then asking them to replace the Class 1, 2 and 3 labels and to think of a story problem with a completely different context, to show that one diagram could represent a variety of different problems.

ASSESSMENT CHECKPOINT Question **4** will allow you to assess which children have a firm understanding of interpreting multi-step word problems. Look in particular at how they have shown their workings.

ANSWERS Answers for the **Practice** part of the lesson can be found in the *Power Maths* online subscription.

Reflect

WAYS OF WORKING Independent thinking

IN FOCUS This section asks children to reflect on how they decide what sort of bar diagram is needed to work out the answer to a problem. They should explain how they know how many bars the bar model needs to have to accurately reflect the problem and to show what the answer will be.

ASSESSMENT CHECKPOINT Look for children linking the pieces of information in a story problem to each bar of a bar model. It may help for children to give an example of a story problem in their explanation, and to describe how to show this on a bar model.

ANSWERS Answers for the **Reflect** part of the lesson can be found in the *Power Maths* online subscription.

After the lesson

- How did children approach **Practice** question **4**?
- Could they create their own story problem?
- Will you need to run any extra intervention in which children solve more problems like these?

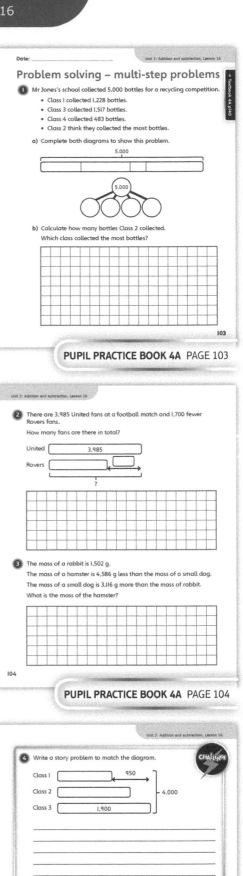

PUPIL PRACTICE BOOK 4A PAGE 103

PUPIL PRACTICE BOOK 4A PAGE 104

PUPIL PRACTICE BOOK 4A PAGE 105

End of unit check

> Don't forget the unit assessment grid in your *Power Maths* online subscription.

WAYS OF WORKING Group work adult led

IN FOCUS This end of unit check will allow you to focus on children's understanding of addition and subtraction and whether they can apply their knowledge to solve problems.

- Look carefully at the answer that children give for question ⑤. It will tell you if they understand how to visually represent and then solve a problem.
- Encourage children to think through or discuss this section before writing their answer in **My journal**.

ANSWERS AND COMMENTARY

Children who have mastered the concepts in this unit should be secure with adding and subtracting 1s, 10s, 100s, 1,000s and adding two 4-digit numbers using the column method. They should be confident subtracting two 4-digit numbers using the column method and be able to use a range of mental addition and subtraction strategies. Children can also find equivalent difference. Children will be able to estimate answers to additions and subtractions, check their strategies and apply knowledge to solve addition and subtraction problems.

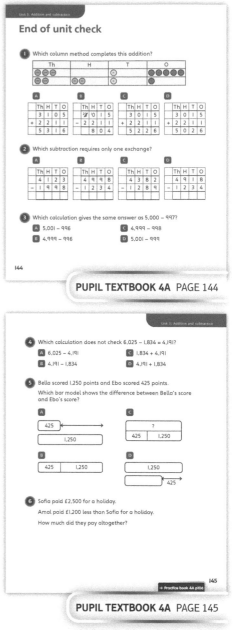

PUPIL TEXTBOOK 4A PAGE 144

PUPIL TEXTBOOK 4A PAGE 145

Q	A	WRONG ANSWERS AND MISCONCEPTIONS	STRENGTHENING UNDERSTANDING
1	C	A suggests that children have misread the place value grid. B is a subtraction rather than an addition. D suggests that children think that $0 + 2 = 0$.	Give practical support with place value to strengthen understanding. Using place value grids may help.
2	D	A indicates children are insecure with identifying exchanges and B that they do not understand the term. C suggests children have not noticed that the exchange from the tens will mean an exchange from the hundreds is needed.	For question 4, run an intervention in which children check answers using the inverse operation.
3	B	Any other answer suggests that children do not understand equivalent difference.	
4	B	A suggests children have misunderstood the calculation. C or D suggest that children do not know how to check an answer using the inverse operation.	Challenge children to match additions, subtractions and word problems with representations such as the bar model.
5	A or D	B and C suggest that children do not understand what 'difference' means.	
6	3,800	Children might have calculated £2,500 – £1,200 and then not worked out the total.	Display the key vocabulary of the unit in your classroom.

My journal

WAYS OF WORKING Independent thinking

ANSWERS AND COMMENTARY

- Question **1**: Children should be able to use their knowledge of rounding to estimate an answer and find out which number is greater than 6,800. They should be able to work this out mentally, without using the column method.

  To complete the first calculation, children will need to use the inverse operation to work out the missing number. $8,634 - 1,849 = 6,785$.

  To complete the second calculation, children will need to calculate $9,000 - 2,026$. They may realise that equivalent difference is a good method here: $8,999 - 2,025 = 6,974$.

- Question **2**: Jamilla scores $4,875 - 3,823 = 1,052$.

  The difference between Aki's score and Lee's score is $8,699 - 4,875 = 3,824$.

  The difference between Aki's score and Jamilla's score is $4,875 - 1,052 = 3,823$.

  Aki's score is closer to Jamilla's score because $3,823 < 3,824$.

  Look for children using diagrams such as bar models to explain their answer and then using column subtraction to work out the differences.

Power check

WAYS OF WORKING Independent thinking

ASK

- *What visual representations and models helped you in this unit?*
- *What do you know now that you did not know at the start of the unit?*
- *What new words have you learnt and what do they mean?*

Power puzzle

WAYS OF WORKING Pair work or small groups

IN FOCUS Use this **Power puzzle** to assess children's problem-solving skills. Can they explain their methods or any strategies that they used?

ANSWERS AND COMMENTARY

Puzzle A: cloud = 1,750   star = 1,250

Puzzle B: heart = 1,050   star = 150   cloud = 1,800   triangle = 600

If children can solve these puzzles, it means they can interpret problems well and use learnt strategies to find a solution. Listen to the explanations of their strategies to check that they have not just guessed a number, but have used reasoning and logic. Encourage children to deepen their understanding by creating their own similar puzzle.

After the unit ⏸

- Which children need further support and how will you provide this support?
- Are children ready for the next unit (Measure – area)? How will you link this unit to finding perimeter?

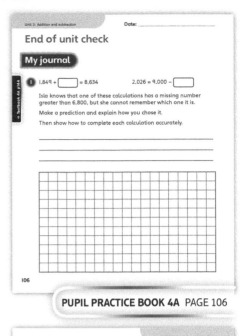

PUPIL PRACTICE BOOK 4A PAGE 106

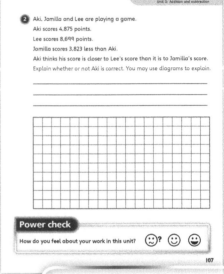

PUPIL PRACTICE BOOK 4A PAGE 107

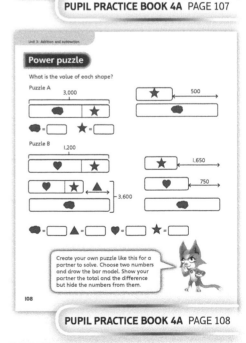

PUPIL PRACTICE BOOK 4A PAGE 108

Strengthen and **Deepen** activities for this unit can be found in the *Power Maths* online subscription.

Unit 4
Measure – area

Don't forget to watch the Unit 4 video!

Mastery Expert tip! 'When I taught this unit, I located examples I could refer to within the school environment – square carpet tiles, square concrete slabs by the school entrance, square ceiling tiles, and so on. I took digital photos and gave children opportunities to devise their own problems based on these real-life examples.'

WHY THIS UNIT IS IMPORTANT

This unit introduces the concept of area to children, giving them a tangible way to measure and compare a shape's size. Until now, children will have been able to say whether a shape is longer or shorter, wider or narrower, and will have been able to measure a shape's length and width. This unit provides them with the tools to measure the space that a shape takes up.

Area is introduced by children using non-standard units and seeing how many of these will fit inside a shape. Children will be made aware of the need for standard units of measurement and will progress to measuring area by counting the number of centimetre squares that fit within a shape. The unit helps children to apply their knowledge in problem-solving and investigative contexts, such as exploring different shapes that all have the same area. Although the relationship between length, width and area is not expounded in this unit, children may recognise informal links between the three concepts from their activities when counting squares. These links will provide a foundation for further development of the concept of area in later years.

WHERE THIS UNIT FITS

→ Unit 3: Addition and subtraction
→ **Unit 4: Measure – area**
→ Unit 5: Multiplication and division (1)

This unit builds on children's understanding of the properties of squares, rectangles and rectilinear shapes. It extends children's basic comprehension of shapes being 'bigger' or 'smaller' than one another and gives them a tangible way of measuring this. Children already know how to measure the distance around a shape and now are taught how to measure the space inside it.

Before they start this unit, it is expected that children:
- understand what is meant by a 2D shape and are able to identify the space inside it
- understand simple properties of squares and rectangles.

ASSESSING MASTERY

Children who have mastered this unit will understand that the area of a 2D shape is the space inside it. They will recognise the importance of using standard units and will be able to confidently measure area by counting squares. They will be able to apply their knowledge to find solutions involving squares, rectangles and rectilinear shapes, including exploring shapes with the same area and comparing shapes based on their areas.

COMMON MISCONCEPTIONS	STRENGTHENING UNDERSTANDING	GOING DEEPER
Children may not recognise the 'conservation of area' (when the squares that form a rectangle are rearranged into different shapes, the area remains the same).	Build links between concrete and pictorial representations by providing squares for children to arrange into the rectilinear shapes shown in their books.	Encourage children to consider half squares and whether they can draw a triangle and work out its area.
Children may assume a shape is automatically larger if it is longer or taller, rather than finding the answer by counting squares.	Give children opportunities to cover areas in different ways. Make links between the area of the shape and the size of the unit of measurement.	Children could be challenged to find the area of rectangular objects using squared paper by drawing around the object, then counting the number of squares within it.

Unit 4: Measure – area

UNIT STARTER PAGES

Use these pages to introduce the concept of the area of 2D shapes to the whole class, checking their understanding of larger or smaller in relation to shapes.

STRUCTURES AND REPRESENTATIONS

Squared paper and square dotted paper overlaid with 2D shapes or counters.

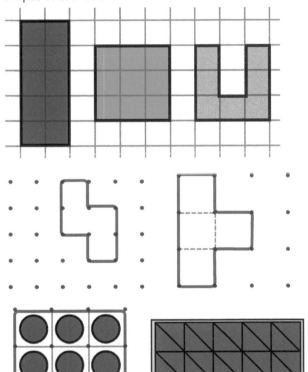

KEY LANGUAGE

There is some key language that children will need to know as part of the learning in this unit:

→ area, space, inside, unit, rows

→ length, width, measure

→ shape, triangle, square, rectangle, trapezium, rectilinear shape, 2D shapes

→ larger, more area, smaller, less area, least area, greater, greatest area

→ right angle

→ counting, subtraction

→ compare, order, size

In this unit we will ...
- ⚡ Learn what 'area' means
- ⚡ Find areas of shapes by counting squares
- ⚡ Draw shapes with different areas
- ⚡ Compare the areas of different shapes

How many small squares fit into this large square?

146

PUPIL TEXTBOOK 4A PAGE 146

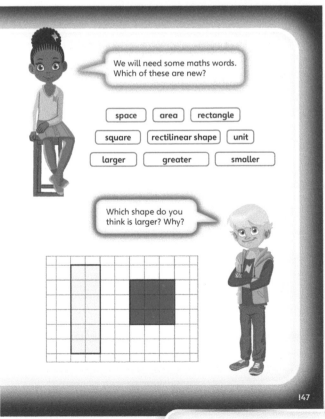

We will need some maths words. Which of these are new?

space   area   rectangle
square   rectilinear shape   unit
larger   greater   smaller

Which shape do you think is larger? Why?

147

PUPIL TEXTBOOK 4A PAGE 147

What is area?

Learning focus

In this lesson, children will be introduced to the concept of the area of a 2D shape. They will measure this by counting non-standard units that fit within squares and rectangles.

Before you teach

- Can you think of any misconceptions that children may have when measuring area?
- How can you use your school environment to introduce and reinforce the concept of area?

NATIONAL CURRICULUM LINKS

Year 4 Measure – area

Find the area of rectilinear shapes by counting squares.

ASSESSING MASTERY

Children can confidently explain that the area of a 2D shape is the space it takes up. Children can express area by measuring this 'space' using non-standard units.

COMMON MISCONCEPTIONS

Children may link a shape's height (or width) to its size, so will assume that a taller shape is larger than a shorter shape. Ask:
- *Which shape takes up the most space? Why do you think this?*

Children may use different types of non-standard units of measure and compare them or may use one type inconsistently (for example, leaving gaps in between them). Ask:
- *Should your units of measure cover the whole shape? Why? What should you remember when choosing the units of measure to use? Does it matter how you arrange the shapes?*

STRENGTHENING UNDERSTANDING

To strengthen understanding, give children simple tasks that involve them exploring area more generally. Ask: *How many children can sit on the mat? How many playing cards will fit on a window ledge? Will a newspaper page fit on a desk top?* Let children cover shapes cut out from paper or 2D shapes with plastic counters. Ask: *How many counters cover this shape? Which of these two shapes needs more counters to cover it?*

GOING DEEPER

Ask children to set up measuring scenarios where they have made a mistake on purpose. For example, showing a book covered in different-sized coins or where the units overlap and/or have gaps between them. Can other children in the class spot the mistakes? Ask: *Is the area the same whatever units you use to measure it? Why not?* Provide a variety of non-standard units and an object to measure.

KEY LANGUAGE

In lesson: space, **area**, triangle, square, rectangle, quadrilateral

Other language to be used by the teacher: 2D shape, space inside, units of measurement

STRUCTURES AND REPRESENTATIONS

2D shapes

RESOURCES

Mandatory: small counters (16 mm preferably), a variety of flat, non-standard units to measure with (flat coloured squares or triangles, playing cards, coins)

Optional: a variety of squares and rectangles to measure (book covers, newspaper pages, paper or card)

 In the eTextbook of this lesson, you will find interactive links to a selection of teaching tools.

Quick recap

Ask children to draw several different rectangles where each one has a different informal property, for example wide, tall, thin or tilted.

Discover

WAYS OF WORKING Pair work

ASK

- Question ① a): *What do you notice about the number of cats on each mat?*
- Question ① b): *What do you notice about the space for the cats?*

IN FOCUS This activity allows children to explore the concept of area through the context of available space within a shape. They compare area by exploring how cramped different-sized shapes might feel when sharing the available space.

PRACTICAL TIPS Try acting out the scenario together. Mark out a small area and safely place three items or people in the area. Repeat with a larger area. Discuss how much space is available in each.

ANSWERS

Question ① a): The mat on the right has more space for the cats, because it has a bigger area.

Question ① b): Ensure children have used counters to fill up the two mats. There should be more counters on the mat on the right-hand side.

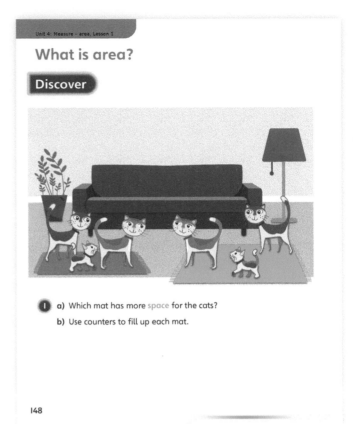

PUPIL TEXTBOOK 4A PAGE 148

Share

WAYS OF WORKING Whole class teacher led

ASK

- Question ① a): *What does it mean to have 'less space' or 'more space'?*
- Question ① a): *Why can these cats be more spread out?*
- Question ① b): *Do these counters cover the mats well?*
- Question ① b): *What could you do to make sure you can fit the most counters possible on each mat?*

IN FOCUS Question ① a) introduces the concept of area as the amount of space inside a given shape. In question ① b), children use non-standard units to make estimates and use informal measurement.

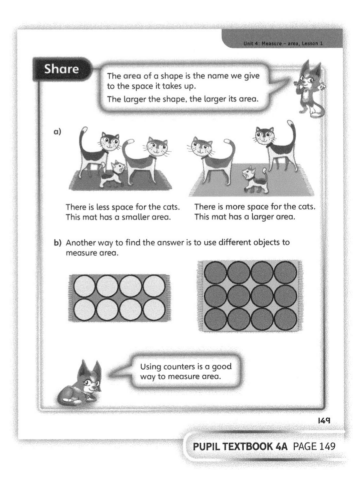

PUPIL TEXTBOOK 4A PAGE 149

Think together

WAYS OF WORKING Whole class teacher led (I do, We do, You do)

ASK

- Question **1**: *Do you think you can fit more than 5 counters in the rectangle? More than 10? More than 20?*
- Question **2**: *Do you think this shape will be easier to measure than the shape in question **1**, or more difficult? Why?*
- Question **3**: *What is the same and what is different about how we have measured the two shapes?*

IN FOCUS In question **1**, children estimate and measure area using non-standard units. In question **2**, they estimate and measure the area using non-standard units for a shape that is not a rectangle. Question **3** gives children the opportunity to explore how different units of measure will give a different degree of accuracy.

STRENGTHEN For questions **1** and **2**, give children opportunities to discuss how to arrange the units of measurement to cover the space inside the shape as best they can. This may involve rearranging the counters to find the best arrangement.

DEEPEN In question **3**, ask children to consider the difference between the two rectangles. Discuss the possible effect on their results of using different-sized units of measurement. Ask: *How can you change things to make a fairer way to compare the areas?*

ASSESSMENT CHECKPOINT Children should understand that 'area' is a measure of the amount of space inside a shape. They should understand which lengths are relevant to measuring an area, and should be able to confidently suggest ways to measure an area using non-standard units of measurement.

ANSWERS

Question **1**: Answers will vary depending on the size of the counters that children are using.

Question **2**: Answers will vary depending on the size of the counters that children are using.

Question **3**: The plastic squares are smaller than the counters. Areas cannot be compared if the unit of measure is not the same.

PUPIL TEXTBOOK 4A PAGE 150

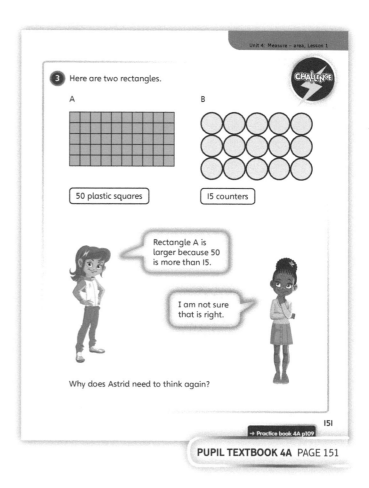

PUPIL TEXTBOOK 4A PAGE 151

Practice

WAYS OF WORKING Independent thinking

IN FOCUS Question ③ is designed to assess children's understanding of the definition of area as the space inside a 2D shape. Question ④ helps elicit the difference between area (the space inside a shape) and length. Question ⑤ underlines the importance of measuring area using a unit of fixed size.

STRENGTHEN Provide children with some physical objects when answering question ③, asking them to use simple non-standard units (such as counters, as they do in question ④) to find each shape's area.

DEEPEN Ask children to estimate and measure the area of different surfaces in the classroom. They should use several different non-standard units to prompt discussion about the changes in estimates required.

ASSESSMENT CHECKPOINT Use question ① to assess whether children can measure area to a reasonable degree of accuracy using non-standard units.

ANSWERS Answers for the **Practice** part of the lesson can be found in the *Power Maths* online subscription.

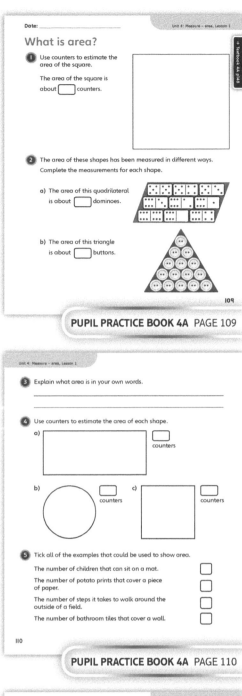

PUPIL PRACTICE BOOK 4A PAGE 109

PUPIL PRACTICE BOOK 4A PAGE 110

Reflect

WAYS OF WORKING Independent thinking

IN FOCUS This question provides an opportunity to check children's methodology. How do they go about looking for an appropriate shape? How do they arrange the counters? Encourage them to explain how they found the area.

ASSESSMENT CHECKPOINT Are children able to define and measure the area of a shape correctly?

ANSWERS Answers for the **Reflect** part of the lesson can be found in the *Power Maths* online subscription.

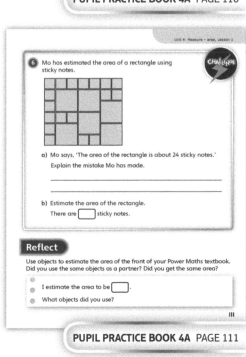

PUPIL PRACTICE BOOK 4A PAGE 111

After the lesson

- Could children have been given more opportunities to explain their reasoning in this lesson?
- Do you feel that children will understand why squares are the standard unit of measurement when they are introduced in the next lesson?

Measure area using squares

Learning focus

In this lesson, children will begin to use squares as a standard unit of measuring the area of squares and rectangles.

Before you teach

- Do children know what the area of a shape is?
- Can you think of squares that occur in your classroom environment that you could use to help children visualise area (such as tiles on a wall)?

NATIONAL CURRICULUM LINKS

Year 4 Measure – area

Find the area of rectilinear shapes by counting squares.

ASSESSING MASTERY

Children can recognise why squares are used as standard units of measurement of area and can confidently apply their knowledge of area by counting squares in shapes made of up to nine squares. Children can write a shape's area as *x* squares (units) and understand what this means.

COMMON MISCONCEPTIONS

Children may confuse the distance around a shape (perimeter) with the space inside it (area). Ask:
- *Can you show me the area of this shape? Now show me the perimeter.*

Children may use standard units inconsistently (for example, leaving gaps between squares) or not understand why squares are a good measure of area (they tessellate, leaving no gaps). Ask:
- *Why do you think using squares is a good way to measure area? Are they better than circles for measuring area? How should the squares be arranged inside the shape?*

STRENGTHENING UNDERSTANDING

To strengthen understanding, give children four flat squares and ask them to rearrange them into different positions. Ask: *What shapes can you make using these squares? What is the area of your shapes?* Repeat with other numbers of squares. Explain that the squares need to tessellate (fit together with no gaps).

GOING DEEPER

Challenge children to demonstrate why squares are an efficient unit for measuring area. Ask: *Can you show why squares are a better measure of area than plastic counters?* Encourage children to consider why squares are better than rectangles as a standard unit of measurement. Ask: *Both squares and rectangles tessellate, which means they fit together with no gaps. Why do you think squares are used as standard units of measurement?*

KEY LANGUAGE

In lesson: units, area, size, shape, measure, squares, rectangle

Other language to be used by the teacher: tessellate, unit of measurement, standard unit, efficient

STRUCTURES AND REPRESENTATIONS

Rectilinear shapes

RESOURCES

Mandatory: squared paper, counters, paper squares

Optional: geoboards, squares cut out from coloured card

 In the eTextbook of this lesson, you will find interactive links to a selection of teaching tools.

Quick recap

Challenge children to estimate the number of counters that would cover the front cover of their maths book. Encourage them to justify their predictions.

Discover

Unit 4: Measure – area, Lesson 2

WAYS OF WORKING Pair work

ASK

- Question **1** a): *Do you remember what the word 'area' means? How might you find the area of these shapes? Do you think squares are a good way to cover a shape? Why or why not?*

IN FOCUS Question **1** a) requires children to visualise a way to cover the space inside the shapes. The fact that both shapes are on geoboards gives them a pictorial indication that the shapes can be split into uniform squares. Observe those children who can see this from the picture and those who will need to cover the area with counters or other non-standard units, as they did in Lesson 1.

PRACTICAL TIPS Use enlarged square dotted paper to replicate the shapes in the picture. Ask children to trace the outline of both shapes. Ask: *Can you see any shapes within these shapes? How many small squares can you see?* Encourage children to draw coloured lines connecting the dots to form squares. Some children may need to number the squares to count them.

ANSWERS

Question **1** a): Shape A has an area of 9 squares (units).
Shape B has an area of 2 squares (units).

Question **1** b): Children should draw a shape with an area of between 2 and 9 squares.

Measure area using squares

Discover

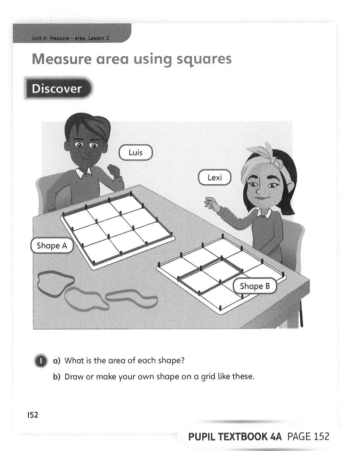

1 a) What is the area of each shape?

b) Draw or make your own shape on a grid like these.

152

PUPIL TEXTBOOK 4A PAGE 152

Share

WAYS OF WORKING Whole class teacher led

ASK

- Question **1** a): *In the last lesson, you measured area by placing counters inside shapes. How is this method the same? How is it different?*
- Question **1** a): *How can you make sure you count all of the squares but do not count any twice or miss any out?*
- Question **1** a): *How would you complete this sentence? 'The shape with the larger area is the one that …'*

IN FOCUS For question **1** a), discuss with children how they might count the squares that fill each shape. Share ideas and provide opportunities for them to try these methods. Children could cover each square with a square of coloured card or colour in each square as it is counted. Using squares cut out from coloured card is a particularly effective way of making the link between the pictorial and the concrete representations.

Share

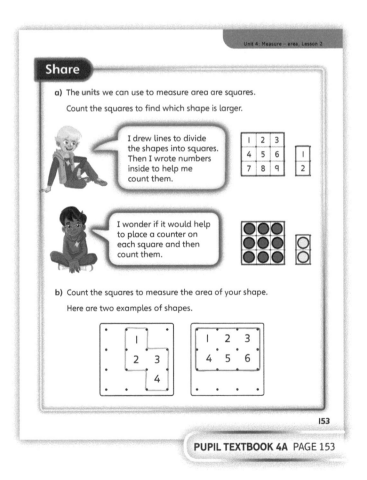

a) The units we can use to measure area are squares.

Count the squares to find which shape is larger.

I drew lines to divide the shapes into squares. Then I wrote numbers inside to help me count them.

I wonder if it would help to place a counter on each square and then count them.

b) Count the squares to measure the area of your shape.

Here are two examples of shapes.

153

PUPIL TEXTBOOK 4A PAGE 153

Think together

Whole class teacher led (I do, We do, You do)

ASK

· Question ❶: *How is the area of these shapes measured? Which shape has the largest area? How do you know? What if you put small 2D squares inside each shape? Would you get the same or a different answer?*

· Question ❷: *What is the same and what is different between the three shapes? How can you work out the area of the final shape?*

IN FOCUS Questions ❶ and ❷ progress from shapes where the inner squares are shown to a shape where children need to visualise the squares that might fit within it. For question ❷, ask: *Which shape has the smallest area? How could you change Shape A so that it has an area that is the same as Shape C?*

STRENGTHEN Provide children with paper squares the same size as the squares shown in the **Textbook**. Invite them to completely cover each shape with squares. Ask: *How many squares cover the shape? What is its area?* This practice will be particularly useful when finding the area of Shape C in questions ❶ and ❷.

DEEPEN In question ❸, ask children to explain the difference in the way the rectangles have been measured. Ask: *What is the same? What is different? Are the rectangles the same size? Why is counting the squares not a helpful way to compare the areas at the moment?* Ensure children recognise that the squares need to be aligned, so there are no gaps between them and they fill the space completely.

ASSESSMENT CHECKPOINT Children should be able to confidently count the number of squares that fill a simple rectilinear shape, giving this value as its area. Children should know how to arrange squares to measure a shape's area efficiently. Question ❸ will raise the issue of what happens when an exact number of squares does not fill a shape completely. You could discuss how Ash could have tried to see if an extra half square would fit in each column, or found a different-sized square that does fit (however, you should emphasise that, to compare areas, you need to use the same square to measure the areas of both shapes).

ANSWERS

Question ❶: The area of Shape A is 5 squares.
The area of Shape B is 4 squares.
The area of Shape C is 6 squares.

Question ❷:

Shape	Area
A	4 squares
B	5 squares
C	7 squares

Question ❸: Various answers are possible. Explanations might mention that the paper squares have not been lined up correctly, there are gaps between them and they do not cover all of the space inside the shape.

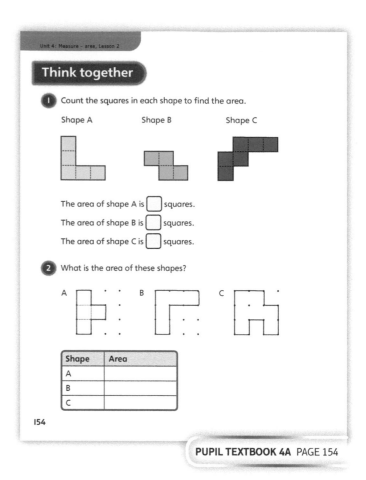

Think together

❶ Count the squares in each shape to find the area.

Shape A   Shape B   Shape C

The area of shape A is ☐ squares.
The area of shape B is ☐ squares.
The area of shape C is ☐ squares.

❷ What is the area of these shapes?

A   B   C

Shape	Area
A	
B	
C	

154

PUPIL TEXTBOOK 4A PAGE 154

❸ Ash has covered these rectangles with small paper squares.

CHALLENGE

These rectangles both have an area of 8 squares. They are the same size.

I do not think this is right. Don't worry, you can learn from your mistake!

What has Ash done wrong?
Explain why you think that this will give the wrong answer.

155

→ Practice book 4A p112

PUPIL TEXTBOOK 4A PAGE 155

Practice

WAYS OF WORKING Independent thinking

IN FOCUS Question **4** addresses one of the main errors children make when measuring area. It reinforces the fact that all units should be arranged so that there are no gaps.

STRENGTHEN Allow children to apply their growing knowledge of area by providing them with squares cut out from coloured card and asking them to make some of the shapes that feature in the questions. Ask: *How many squares did you use? What is the shape's area? Can you move the squares to make the same shape, but in a different position? Has the area changed? Can you move some of the squares to make a completely new shape? Has the area changed?*

DEEPEN In question **6**, children are presented with a sequence of squares. Children can explore the pattern in the areas on several levels: by counting the number of small squares within each larger square and then recognising visually how the pattern continues; by drawing or predicting the next squares in the sequence; and by recognising the numerical pattern (the area increases by the next odd number each time: + 3, + 5, + 7, and so on).

THINK DIFFERENTLY Question **5** provides children with a code to crack. They are asked to work out the area of each shape, then find the relevant letter using the key. When they write the code letter, a phrase will be revealed. Encourage them to find the area of that object (table top) using physical squares.

ASSESSMENT CHECKPOINT Children should be able to find areas of simple shapes by counting squares. Some children may need manipulables, others will be able to count squares in images, and some may visualise the shapes described. They should recognise that a shape is the same size as another if it has the same area.

ANSWERS Answers for the **Practice** part of the lesson can be found in the *Power Maths* online subscription.

Reflect

WAYS OF WORKING Independent thinking

IN FOCUS This question has been designed to assess how children's methodology has progressed over the course of the lesson. Children should now describe how to find area using standard units, by counting squares.

ASSESSMENT CHECKPOINT Look for children who are able to confidently and clearly describe how they can count squares in order to find the area of simple shapes.

ANSWERS Answers for the **Reflect** part of the lesson can be found in the *Power Maths* online subscription.

After the lesson

- Are children confident in finding areas by counting squares?
- Do children recognise why squares are an efficient unit of measurement to use when finding area?

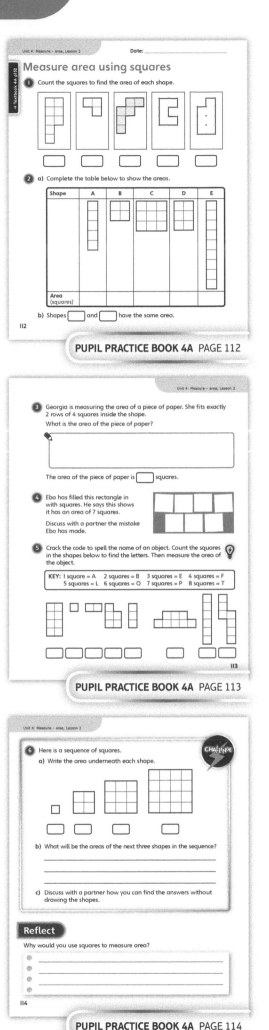

PUPIL PRACTICE BOOK 4A PAGE 112

PUPIL PRACTICE BOOK 4A PAGE 113

PUPIL PRACTICE BOOK 4A PAGE 114

Count squares

Learning focus

In this lesson, children will find areas of more complex rectilinear shapes (including those drawn on squared grids) by counting squares.

Before you teach ⏸

- How will you define the term 'rectilinear' and describe rectilinear shapes (particularly as squares and rectangles are both rectilinear shapes)?
- In your school or classroom environment, are there squares arranged in grids (such as tiles) that you could use to illustrate real-life rectilinear shapes?

NATIONAL CURRICULUM LINKS

Year 4 Measure – area

Find the area of rectilinear shapes by counting squares.

ASSESSING MASTERY

Children can confidently find the area of rectilinear shapes drawn on a square grid by counting squares, write the area as *x* squares (units) and understand what this means. Children can apply their knowledge of area to find areas in different contexts.

COMMON MISCONCEPTIONS

Children may only count the squares around the edge or outside of a shape, rather than looking at the whole area. Ask:
- *Have you counted all the squares inside the shape? Can you use counters to check your answer?*

STRENGTHENING UNDERSTANDING

Provide children with concrete opportunities to recognise and measure the area of rectilinear shapes around school. Use masking tape around floor tiles in a corridor or use chalk around square paving slabs to show simple rectilinear shapes. Ask: *Which part of this shape is its area? What is the area of this shape?* Explore different ways of showing that each square has only been counted once (such as placing a sticky note inside it, drawing a chalk dot inside it, standing inside it, etc.). The activity could be framed as an area-based scavenger hunt.

GOING DEEPER

Set children the challenge of designing their own ideal bedroom or creating a classroom plan on squared paper. Limit the objects within their plans to rectilinear shapes. Ask them to devise questions based on their plans (for example: *What is the area of …? Which is larger, the … or the …? How much free space is in the room?*). By writing down the expected answers, children are applying their knowledge of area in a real-life context.

KEY LANGUAGE

In lesson: area, space, squares, rows, rectilinear shape, right angle, rectangle, counting, subtraction

Other language to be used by the teacher: 2D shape, units of measurement (or 'units'), plan

STRUCTURES AND REPRESENTATIONS

Rectilinear shapes

RESOURCES

Mandatory: squared paper, flat plastic shapes

Optional: counters, sticky notes, chalk, masking tape, rectangles cut out from coloured card

 In the eTextbook of this lesson, you will find interactive links to a selection of teaching tools.

Quick recap ↻

Ask children to find the area of some given rectangles drawn on squared paper, by counting the squares.

Discover

Unit 4: Measure – area, Lesson 3

WAYS OF WORKING Pair work

ASK

- Question ❶: *Which part of each plan do you need to measure to find the area of the bed? Or the empty space? Why do you think the bedroom plans have been drawn on squared paper?*

IN FOCUS In both parts of question ❶, children may notice that there are different strategies they can use to count squares to find the area. Some children may use their knowledge of arrays by counting the number of rows and columns. In question ❶ b), children may note that the space in each bedroom is the same as the total area minus the size of the bed.

PRACTICAL TIPS Provide children with flat plastic squares. Ask them to arrange these to make the same shape as each bed. Link these concrete examples of rectangles with the images in the **Textbook**. Children could also make shapes to represent other items found in the bedroom, such as a chest of drawers or a table.

ANSWERS

Question ❶ a): Kate's bed has an area of 10 squares.
Aki's bed has an area of 12 squares.

Question ❶ b): Kate has 26 squares of empty space.
Aki has 23 squares of empty space.

Count squares

Discover

❶ a) What is the area of each bed?

b) What is the area of empty space Kate and Aki each have in their bedroom?

156

PUPIL TEXTBOOK 4A PAGE 156

Share

WAYS OF WORKING Whole class teacher led

ASK

- Question ❶ a): *What can you do to find the area of the different shapes? How many squares are there in one row? How many rows are there? How does this information help?*

IN FOCUS Encourage children to model how they would find each rectangle's area accurately, without double-counting or missing any squares. When considering question ❶ b), challenge children to consider the relative sizes of the shapes. Ask: *Does the fact that Aki's bed is larger mean that the space in his room will always be less? Does it depend on something else?* Some children may find the answer in different ways. For example, they can find the total number of squares in the room and subtract the size of the bed.

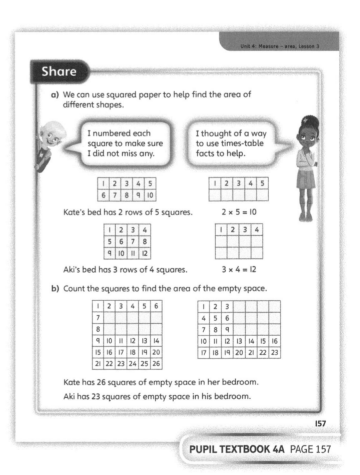

Share

a) We can use squared paper to help find the area of different shapes.

> I numbered each square to make sure I did not miss any.

> I thought of a way to use times-table facts to help.

Kate's bed has 2 rows of 5 squares.

$2 \times 5 = 10$

Aki's bed has 3 rows of 4 squares.

$3 \times 4 = 12$

b) Count the squares to find the area of the empty space.

Kate has 26 squares of empty space in her bedroom.
Aki has 23 squares of empty space in his bedroom.

157

PUPIL TEXTBOOK 4A PAGE 157

Think together

WAYS OF WORKING Whole class teacher led (I do, We do, You do)

ASK

- Question **1**: *Which shape looks like it has the largest area? Is there more than one way to find the area of these shapes?*
- Question **2**: *Compare these shapes with squares and rectangles. What is the same? What is different? Is finding the area of these shapes the same as finding the area of a square or a rectangle? How?*

IN FOCUS Question **2** introduces children to the concept that area is not limited to squares and rectangles. Encourage them to describe each shape in their own words. Explain that if a rectilinear shape looks like two rectangles put together, they could work out the area of the whole shape by adding the areas of the rectangles together. Give them opportunities to draw some rectilinear shapes on the board.

STRENGTHEN To ensure that children are counting all the squares inside each shape, encourage them to place counters on each square as they count them. Alternatively, ask children to draw out the shapes on squared paper and number each square as they count them.

DEEPEN Question **3** involves finding the area of a rectangle that is partly concealed, so that not all of the squares inside can be counted. Children should use the fact that the number of squares in each row or column is the same, regardless of whether they are covered or not.

Deepen this concept by asking children to place an opaque shape over a square grid and then challenging them to work out the area of the grid.

ASSESSMENT CHECKPOINT At this point, children should be able to confidently find the area of these larger rectilinear shapes made up of squares. They should do so either by counting the visible squares inside the shape or by using the squared paper around the shape to visualise the number of squares it covers. In question **2**, children may notice that the areas can be found as the sum of two rectangles, in the form $a \times b + c \times d$.

ANSWERS

Question **1**: Rectangle A = 20 squares
Rectangle B = 18 squares
Rectangle C = 18 squares

Question **2**: The area of Shape A is 30 squares.
The area of Shape B is 30 squares.

Question **3**: The area of this shape is 40 squares.

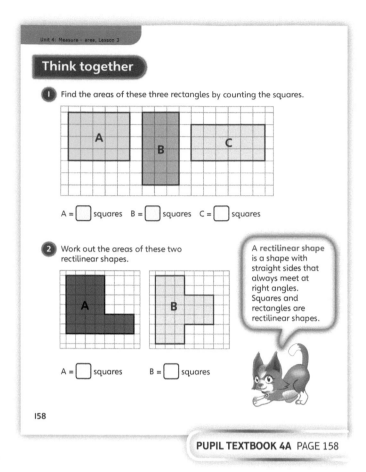

PUPIL TEXTBOOK 4A PAGE 158

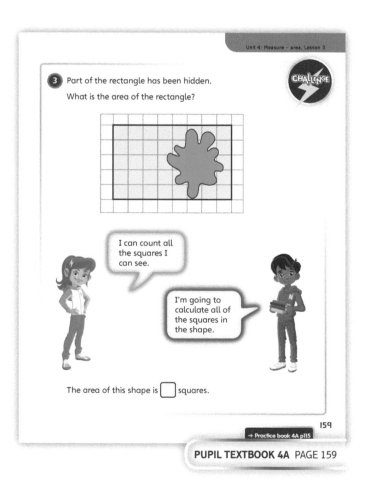

PUPIL TEXTBOOK 4A PAGE 159

Practice

WAYS OF WORKING Independent thinking

IN FOCUS Question ② shows two rectangles adjacent to each other. Children are guided to find the area of each individual rectangle to find the total area. This is an important model to help children recognise that rectilinear shapes can be split into smaller squares and rectangles to find their area. If time allows, ask children to describe how some of the rectilinear shapes in question ① could be split into smaller squares and rectangles too. Question ④ gives children the opportunity to find the area of a shape when some of the squares inside are covered.

STRENGTHEN Encourage children to write on the grids in their **Practice Books** and devise a way of ensuring that they count the number of squares correctly. They might choose to do this by marking each square as it is counted (tick, colour or dot) or by writing numbers in each square. As well as ensuring that their counting is accurate, this will also help children to visualise the area of the shape.

DEEPEN Question ⑤ is a challenge requiring several steps. Lead children though the process. Ask: *How can you find the number of squares around the farmhouse? (Count: 15) This total area is 15 squares, and there are 5 fields, so how many squares will each field be? (3) Which different shapes could you draw with an area of 3 squares? Can you think of a way to divide the land so that all the fields are the same shape?*

ASSESSMENT CHECKPOINT Children should be able to find the area of any rectilinear shape drawn on a grid, including examples where rectilinear shapes have been split into separate rectangles or squares.

ANSWERS Answers for the **Practice** part of the lesson can be found in the *Power Maths* online subscription.

Reflect

WAYS OF WORKING Pair work

IN FOCUS The **Reflect** part of the lesson prompts children to use counting strategies to reason the total area of a shape when some of the squares on the grid are obscured.

ASSESSMENT CHECKPOINT Assess whether children can justify their count by reasoning about the number of squares on the grid.

ANSWERS Answers for the **Reflect** part of the lesson can be found in the *Power Maths* online subscription.

After the lesson ⏸

- Do children completely understand the concept of a shape's area (as opposed to other measurements: length, width, perimeter)?
- Were children able to find the area of simple composite shapes made up of squares and rectangles?

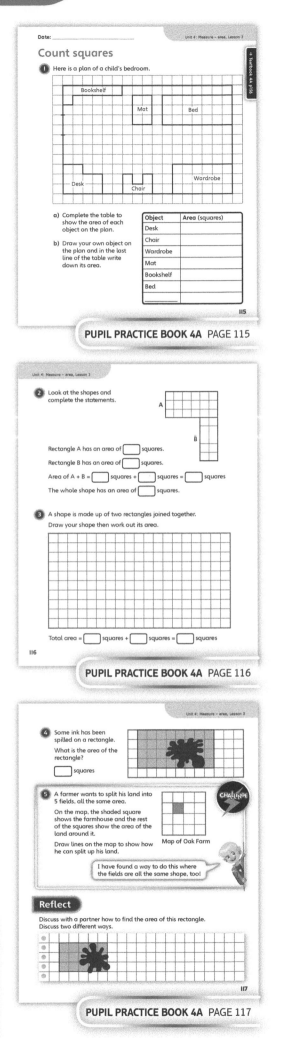

PUPIL PRACTICE BOOK 4A PAGE 115

PUPIL PRACTICE BOOK 4A PAGE 116

PUPIL PRACTICE BOOK 4A PAGE 117

Make shapes

Learning focus

In this lesson, children will be given opportunities to apply their understanding of the concept of area by making shapes with given areas.

Before you teach

- How will you ensure children recognise that, although a shape looks different to another, they may have the same area?
- What opportunities will you give children to use reasoning during the lesson?

NATIONAL CURRICULUM LINKS

Year 4 Measure – area

Find the area of rectilinear shapes by counting squares.

ASSESSING MASTERY

Children can apply their knowledge of area by making and drawing rectilinear shapes with a given area (expressed in terms of number of squares).

COMMON MISCONCEPTIONS

Children may not recognise the 'conservation of area' (for example, that when a rectangle is rearranged into a different shape, it still has the same area as the original rectangle). Children may also see a long, thin combination of squares as having a different area to, for example, the same number of squares arranged in a block. Ask:

- *What is the area of this shape? Can you rearrange the squares into a different shape? What is its area now? Is that the same or a different area to the first shape?*

STRENGTHENING UNDERSTANDING

Take children to an area of the school where there are squares arranged in a grid (such as tiles by the sink, concrete slabs around the playground or carpet tiles). Let them mark the outline of different shapes with masking tape or chalk, asking them to make a shape with a given area. Concentrate on whether the shapes have the given area, rather than whether any are repeated or whether all possibilities have been found. Take digital photos of the shapes that children make and use them as a display.

GOING DEEPER

Challenge children to find and draw five or more different rectangles or rectilinear shapes that each have an area of exactly 12 squares.

KEY LANGUAGE

In lesson: shape, area, rectilinear shape, rectangle, square(s), reflection, rotation

Other language to be used by the teacher: duplicate, rotate, reflect, flip

STRUCTURES AND REPRESENTATIONS

2D shapes

RESOURCES

Mandatory: 1 cm paper squares or counters, squared paper, square dotted paper

Optional: chalk, masking tape, square tiles

 In the eTextbook of this lesson, you will find interactive links to a selection of teaching tools.

Quick recap

Have a discussion about the number of whiteboards that would cover the area of the school playground or hall. Encourage children to explain their reasoning.

Discover

Pair work

ASK

- Question ❶: *How has the patio in the picture been made? What is the area of the patio that is shown?*
- Question ❶: *What is the rule for how the patio slabs must fit together?*

IN FOCUS Children are given a practical problem to solve. This requires them to generate different rectilinear shapes by joining a given number of square units along their edges.

PRACTICAL TIPS Ask children whether they have any experience of arranging squares to make a shape; perhaps they have seen adults making a patio or tiling a wall. They may have played a board game where they had to place squares next to each other to make different shapes.

ANSWERS

Question ❶ a): Answers will vary. Ensure children have drawn a rectilinear shape with 4 squares.

Question ❶ b): Answers will vary. Ensure children have drawn a rectilinear shape with 5 squares.

PUPIL TEXTBOOK 4A PAGE 160

Share

Whole class teacher led

ASK

- Question ❶: *How would you describe each of these shapes? Is it true that there can be lots of different shapes that all have the same area? Why?*
- Questions ❶ b): *Can you think of any other arrangements to try?*

IN FOCUS In question ❶ a), children make a rectilinear shape with a given area expressed in terms of squares. In question ❶ b), they make another, larger, rectilinear shape with a given area. There is more than one possible answer and children should be encouraged to work systematically to try to find them all.

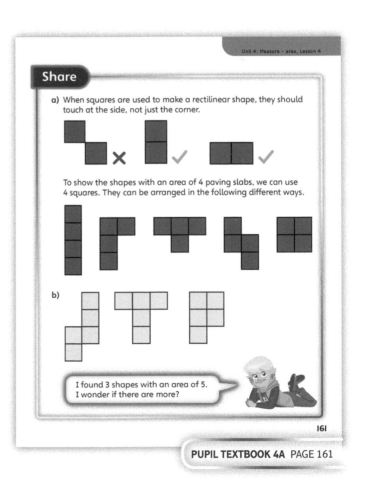

PUPIL TEXTBOOK 4A PAGE 161

Think together

WAYS OF WORKING Whole class teacher led (I do, We do, You do)

ASK

- Question **1**: *Can you find more than three solutions?*
- Question **2**: *How can you be sure you have drawn a square?*
- Question **3**: *What is the same and what is different about each of the shapes?*

IN FOCUS In question **1**, children create their own rectilinear shape with a given area expressed in terms of squares. In question **2**, they create specific shapes with given areas, exploring how many possible solutions they can find. Question **3** requires children to explore the properties of rectilinear shapes and to identify which shape has a given area.

STRENGTHEN Ask children to use square tiles or 2D shapes to make shapes with given areas expressed as squares.

DEEPEN Challenge children to find all the possible solutions for shapes with given properties and area. Ask: *How many different rectangles can you find with an area of 16 squares?*

ASSESSMENT CHECKPOINT In question **2**, assess whether children can explain why there is only one possible square solution, but more than one possible rectangle solution.

ANSWERS

Question **1**: Answers will vary. Ensure children have drawn a shape with an area of 8 squares.

Question **2**: The square is 5 × 5 squares.
A rectangle of 24 squares can have dimensions: 12 × 2, 8 × 3, 6 × 4 or 1 × 24.

Question **3**: Holly can make shape C, because this is the only shape that is rectilinear and has an area of 8 squares.

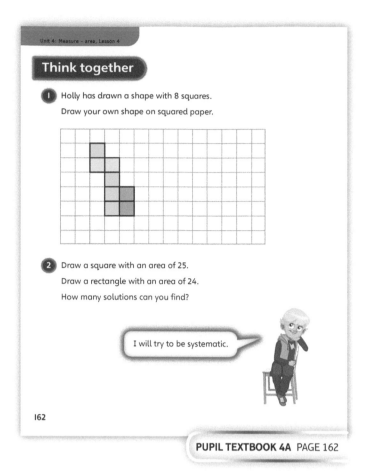

PUPIL TEXTBOOK 4A PAGE 162

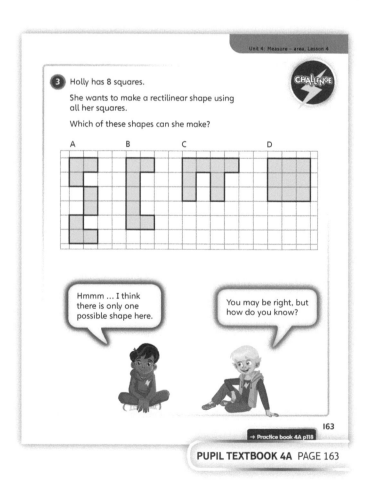

PUPIL TEXTBOOK 4A PAGE 163

Practice

WAYS OF WORKING Independent thinking

IN FOCUS In questions ❶ and ❷, children draw rectilinear shapes of a given area where the area is expressed in terms of squares. In question ❸, children reason about the area of given shapes, where the number of squares inside each shape cannot be counted. Questions ❹ and ❺ are creative problem-solving questions where children use given criteria to make their own shapes.

STRENGTHEN Question ❷ requires children to visualise whether certain shapes can be made with a given area. Provide children with four paper or cardboard squares, so they can physically explore the shapes that can be made before drawing them. In question ❹, children could just draw their initials.

DEEPEN Question ❹ asks children to investigate the areas of their names. Encourage children to draw the letters using squares (no diagonal lines). Most names will contain 'holes'. Ask: *Is the hole part of the area of the shape? Should you count it or not? Do any letters have the same area? Do they look the same size?*

ASSESSMENT CHECKPOINT Use questions ❶ and ❷ to assess whether children can draw rectilinear shapes where the given area is expressed in terms of squares.

ANSWERS Answers for the **Practice** part of the lesson can be found in the *Power Maths* online subscription.

Reflect

WAYS OF WORKING Independent thinking

IN FOCUS Children are asked to devise three rules to follow when making different rectilinear shapes out of the same number of squares. Their answers may differ, but should demonstrate an understanding of how to make shapes with a given area.

ASSESSMENT CHECKPOINT Look for children who write appropriate 'rules' that clearly describe the methodology of making shapes with a given area.

ANSWERS Answers for the **Reflect** part of the lesson can be found in the *Power Maths* online subscription.

After the lesson ⏸

- Are children able to confidently make shapes with a given area?
- What other cross-curricular opportunities can be provided to encourage children to work systematically?

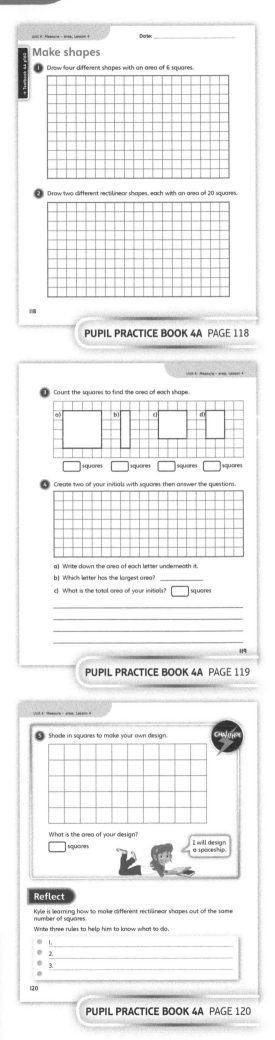

PUPIL PRACTICE BOOK 4A PAGE 118

PUPIL PRACTICE BOOK 4A PAGE 119

PUPIL PRACTICE BOOK 4A PAGE 120

Compare area

Learning focus

In this lesson, children will learn how to compare shapes according to their areas.

Before you teach

- What misconceptions have children shown when comparing numbers? Could they make the same mistakes when comparing areas?
- How can you explain to your class the concept of winning a game by having the largest area? Can you think of any games where this is the case?

NATIONAL CURRICULUM LINKS

Year 4 Measure – area

Estimate, compare and calculate different measures.

ASSESSING MASTERY

Children can confidently compare the area of rectilinear shapes by first measuring their area accurately by counting squares and then comparing both values to see which is the larger. Children can apply this skill to order several shapes according to their areas.

COMMON MISCONCEPTIONS

Children may compare a shape's area visually, by the way it looks rather than by counting squares. As area is dependent on two dimensions (length and width), children may find it difficult to compare shapes when both dimensions differ. Ask:
- *What do you need to do to work out which shape is larger?*

STRENGTHENING UNDERSTANDING

Play a variation on the game in the **Discover** section. Display a grid on the board and divide the group into a number of teams, each team with different patterned squares. They take it in turns to put a patterned square onto the board. They can try to build a shape, but can also block each other by placing their patterned square where other teams might want to go. When all the board is covered, ask:
- *Which shape has the largest area? How could you order the shapes in terms of area?*

Draw the shapes away from the board and show them in order.

GOING DEEPER

Challenge children to devise their own area-based games that include both measurement of the area by counting squares and comparison of area. Children could then play these games to practise these skills.

KEY LANGUAGE

In lesson: greater, area, compare, order, size

Other language to be used by the teacher: multiply, total, subtract, count, rectangle, rectilinear shape, grid, measure, accurately

STRUCTURES AND REPRESENTATIONS

2D shapes

RESOURCES

Optional: a large grid, patterned squares (or pens to colour in the squares), manipulatives (counters, small squares)

 In the eTextbook of this lesson, you will find interactive links to a selection of teaching tools.

Quick recap

Ask children to draw several different rectangles where each one has a different informal property, for example wide, tall, thin or tilted. Then ask them to carefully measure the area of each one.

Discover

Pair work

ASK

- Question **1** a): *Can you predict who is winning the game just by looking? Why do you think this? Which shape is taller? Does this mean it will have the larger area?*
- Question **1** b): *How can you compare the areas?*

IN FOCUS Both parts of this question encourage children to compare the areas of two different shapes. They will already have plenty of experience of looking at two shapes and choosing the one that looks larger. Now they are required to show the one that is larger, finding each area by counting squares and then comparing the values.

PRACTICAL TIPS Play a similar game to the one shown in **Discover**. Display a small grid on the board and divide the group into two teams. They take it in turns to colour any square on the board. If they manage to 'trap' squares of their opponent's colour between two squares of their own colour, they can change the trapped squares into their colour (this is the same concept as the game 'Othello'). After children have had several turns, pause the game and ask: *Which shape has the largest area?* Only include rectilinear shapes.

ANSWERS

Question **1** a): 17 > 16, so Olivia is winning the game.

Question **1** b): The total area of the board is:
10 × 10 = 100 squares.
100 − (17 + 16) = 100 − 33 = 67.
The area of the board that is not covered is larger because 67 > 33.

Share

WAYS OF WORKING Whole class teacher led

ASK

- Question **1** a): *Can you tell who is winning just by looking? Why or why not? How can you check which shape has the larger area? If Danny has beaten Olivia at the game, what can you tell me about the area of Danny's shape?*
- Question **1** b): *One way to find the number of white squares is to count all the white squares. Is there another way to find the same answer?*

IN FOCUS Encourage children to work out question **1** b) using both methods to check their answer. Firstly, counting all the white squares, then working out that there are 10 × 10 = 100 squares on the board and subtracting the total number of coloured squares. Discuss which method they found easier.

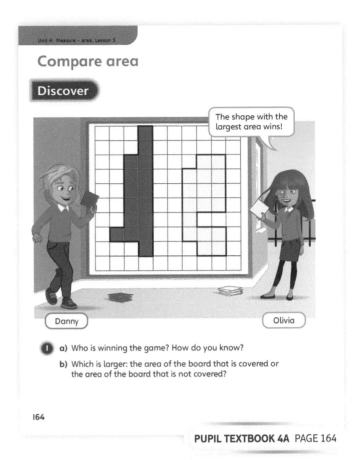

PUPIL TEXTBOOK 4A PAGE 164

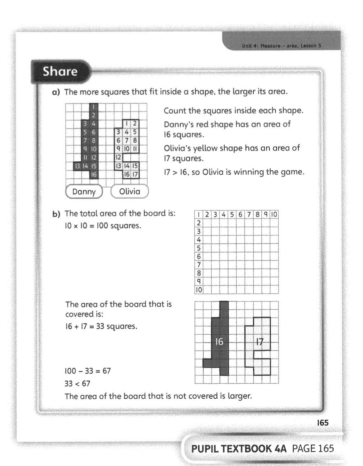

PUPIL TEXTBOOK 4A PAGE 165

Think together

Whole class teacher led (I do, We do, You do)

ASK

- Question **1**: *What is the only way to compare areas accurately?*
- Question **2**: *Are the squares inside the middle of the letter a part of its area? Why or why not? What steps do you need to go through to put these letters into the correct order of size?*

IN FOCUS Questions **1** and **2** take children through the steps needed to compare the area of rectilinear shapes. First they should find the areas by counting squares. Then they should compare the values to see which is larger. In question **2** b), children are expected to use their comparison to write the areas in order.

STRENGTHEN For questions **1** and **2**, give children manipulatives to place on the shapes (one for each square). They should count these and place each set of manipulatives in a line. The longer line shows a concrete representation of the greater area. Move on from this to examples where children simply locate the values on a number line. Ask: *Which number is larger? How do you know?*

DEEPEN In question **3**, discuss the mistake Astrid has made. Ask: *Is this a common mistake to make?* Remind children that we all make visual judgements about measurements all the time. Ask: *What is the only way to make sure you compare the areas accurately?* Ensure children can also explain how to find the area of both shapes.

ASSESSMENT CHECKPOINT Children should understand that they cannot rely on visual cues to compare shapes, and that the only accurate way is to find their areas and then compare the values.

ANSWERS

Question **1**: Area of A = 18 squares
Area of B = 19 squares
19 > 18
Shape B has the larger area.

Question **2** a): Letter g = 22 squares
Letter s = 21 squares
Letter t = 17 squares

Question **2** b): t, s, g

Question **3**: Astrid's shape has 18 squares; Flo's shape has 20 squares, so Flo is the winner. Children should mention the fact that the only way to compare areas is to count the number of squares and then compare these numbers. A shape can be taller and/or wider, but still have a smaller area.

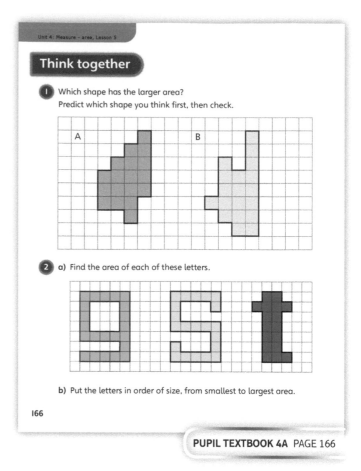

Think together

1 Which shape has the larger area?
Predict which shape you think first, then check.

2 a) Find the area of each of these letters.

b) Put the letters in order of size, from smallest to largest area.

166

PUPIL TEXTBOOK 4A PAGE 166

3 The shape with the greater area wins. Who has won the game and why?

CHALLENGE

Flo's shape

Astrid's shape

I win! My purple shape is a lot taller than your rectangle, so the area must be greater.

I think we need to do something else to be able to compare areas.

167

→ Practice book 4A p121

PUPIL TEXTBOOK 4A PAGE 167

Practice

WAYS OF WORKING Independent thinking

IN FOCUS Question **3** shows a picture sequence. Ask children what they notice. Direct their attention to what is happening to the left-hand shapes as the sequence progresses, then to the right-hand shapes. Ask: *How many squares are added to this shape each time? Where have they been drawn?* When the sequence starts, the left-hand shape in each pair has the greater area, but this changes at part c) (same area) and reverses at part d). Ask: *If you continued the sequence, would the shape on the left ever be larger again?*

STRENGTHEN To help children count accurately in the larger shapes, they could write the total of each line as they go, then add the lines to find the overall area of the shape.

DEEPEN Question **4** involves reasoning about the area of shapes. Deepen this by asking children to draw a square on squared or square dotted paper, and then to find its area. Now challenge them to find and draw a rectangle with an area that is exactly 1 square less than that of the square that they drew first.

ASSESSMENT CHECKPOINT Children should be working confidently to find areas by counting squares and then comparing and ordering those areas. Children should be able to use reasoning to explain their answers and make generalisations.

ANSWERS Answers for the **Practice** part of the lesson can be found in the *Power Maths* online subscription.

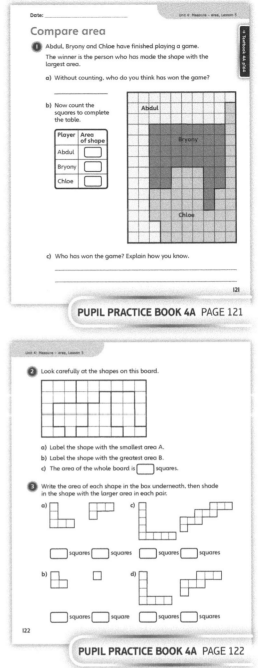

PUPIL PRACTICE BOOK 4A PAGE 121

Reflect

WAYS OF WORKING Independent thinking

IN FOCUS The **Reflect** part of the lesson prompts children to understand that, even though a rectangle may be longer than a square, it may have a smaller area. Area is a measure of the space that the shape takes up, not the length of its sides.

ASSESSMENT CHECKPOINT Assess whether children can explain why the area of a square may be greater or less than the area of a rectangle.

ANSWERS Answers for the **Reflect** part of the lesson can be found in the *Power Maths* online subscription.

PUPIL PRACTICE BOOK 4A PAGE 122

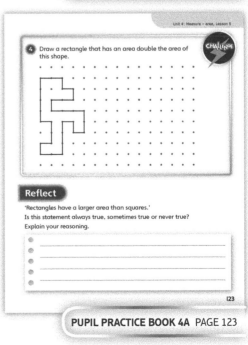

PUPIL PRACTICE BOOK 4A PAGE 123

After the lesson ⏸

- How confident are children when comparing the area of rectilinear shapes?
- Are children now less reliant on comparison by sight and do they understand why it is important to measure before comparing the size of shapes?

End of unit check

Don't forget the unit assessment grid in your *Power Maths* online subscription.

WAYS OF WORKING Group work adult led

IN FOCUS

- Question ❶ assesses children's ability to define the area of a shape.
- Question ❷ assesses children's ability to find the area of a rectangle drawn on squared paper by counting squares within it.
- Question ❸ assesses children's ability to find the area of solid squares and rectangles drawn on squared paper by using the squares around them to work out the number of squares within them.
- Question ❹ assesses children's ability to compare the area of rectilinear shapes.
- Question ❺ assesses children's ability to make different shapes with a given area.
- Question ❻ is a SATs-style question, assessing children's ability to identify a shaded area as the difference between the areas of the larger rectangle and the smaller square.

ANSWERS AND COMMENTARY Children who have mastered the concepts in this unit will be able to define the term 'area' and calculate the area of rectilinear shapes by counting the squares within them. They will be able to apply their knowledge to find solutions involving solid shapes drawn on grids, comparing areas and making different shapes with a given area.

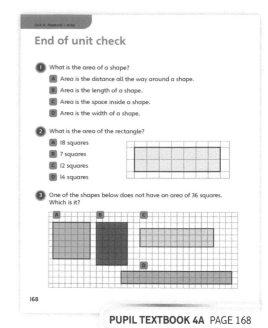

PUPIL TEXTBOOK 4A PAGE 168

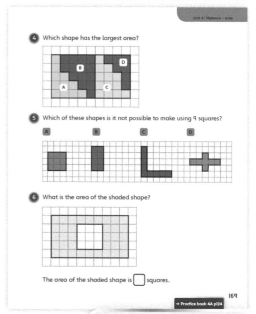

PUPIL TEXTBOOK 4A PAGE 169

Q	A	WRONG ANSWERS AND MISCONCEPTIONS	STRENGTHENING UNDERSTANDING
1	C	B or D suggest children link area with one dimensional measurements (whether a shape is longer or wider).	Some of the misconceptions children may have about area will be a result of not understanding what the area of a shape is. Strengthen their understanding of the general concept of area being the space inside a shape by providing opportunities to use concrete resources (such as flat plastic squares, squares cut out of paper, acetate squared grids) to measure the number of squares that fit inside a shape. Allow them to draw rectilinear shapes on squared paper, then cover the shapes with coloured paper squares.
2	D	A is the perimeter; B suggests children counted the columns; C suggests inaccurate counting.	
3	B	A, C and D suggest children are unsure how to use the squares around a solid shape to help derive its area.	
4	B	A, C and D suggest children have either miscounted squares or compared areas incorrectly.	
5	B	A, C or D suggest children have not visualised the possible shapes correctly.	
6	36 squares	45 squares is the area of the shaded rectangle without removing the inner white area.	

My journal

Independent thinking

ANSWERS AND COMMENTARY Each shape should be drawn on the squared dotted paper provided and should have an area of 12 squares.

Possible answers are:
- rectangles with dimensions 1 × 12, 4 × 3 or 3 × 4.
- various other rectilinear shapes (not including squares as it is not possible to make a square using 12 squares).

To help children explain how to choose their measurements, ask:
- *Draw a rectangle. Does it have an area of 12 squares? If not, how could you adjust it so that it does? If so, how could you adjust it so that it still has an area of 12 squares?*
- *Is it possible to draw a square with an area of 12 squares? How do you know?*

Power check

WAYS OF WORKING Independent thinking

ASK
- *What did you know about area before you began this unit?*
- *What do you know now?*
- *Do you think you would be able to find the area of any rectilinear shape drawn on squared paper on your own?*

Power play

WAYS OF WORKING Pair work or small groups

IN FOCUS Use this **Power play** to see if children can explore combinations of the given rectilinear shapes to make two rectangles with different areas. Children should be able to use their knowledge of the properties of rectangles to derive both shapes before calculating their areas.

ANSWERS AND COMMENTARY

a) The two rectangles formed should have the dimensions 3 × 7 and 5 × 4 and look as follows:

b) The areas of the chocolate bars are 21 squares and 20 squares.

c) Children should mention the fact that the bar containing 21 squares has a larger area (whether they would choose this bar depends on whether or not they like chocolate!).

Completing the **Power play** shows that they understand the properties of rectangles and can find their area by counting the squares. If they are unable to complete the question, check whether they have simply been unable to make the rectangles or, having made them, are unsure how to calculate or compare the areas.

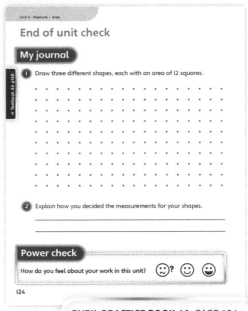

PUPIL PRACTICE BOOK 4A PAGE 124

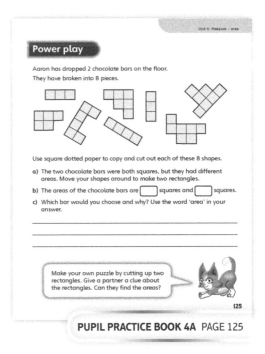

PUPIL PRACTICE BOOK 4A PAGE 125

After the unit ⏸

- What will you do differently next time you teach this unit?
- How did children respond to the materials and approaches used? Were they adequately (or excessively) challenged by them?

Strengthen and **Deepen** activities for this unit can be found in the *Power Maths* online subscription.

Unit 5
Multiplication and division ❶

Mastery Expert tip! 'This unit is a superb opportunity to encourage all children to be able to demonstrate rapid recall of times-tables. It is important to ensure that children know the multiplication facts and related division facts, but also that they are able to explain them. Children should be encouraged to use visual representations such as arrays and number lines.'

Don't forget to watch the Unit 5 video!

WHY THIS UNIT IS IMPORTANT

This unit is important because it focuses on learning multiplication and division facts – a core part of maths at Key Stage 2. Children explore multiplication and division, looking first at multiplying and dividing by 3, 6, 9, 7, 11 and 12, and then at multiplying and dividing by 0 and 1, the understanding of which is key to children's mastery of this unit. This unit encourages children to use visual representations to tackle multiplication and division questions. Mastering this unit will certainly have a positive impact on other areas of mathematics such as fractions, decimals and percentages.

WHERE THIS UNIT FITS

→ Unit 4: Measure – area
→ **Unit 5: Multiplication and division (1)**
→ Unit 6: Multiplication and division (2)

This unit builds upon the previous work children have done on multiplication and division from Year 3, where children learnt how to multiply by equal grouping and to divide using sharing. This unit also builds upon previous work children have done on addition and subtraction. It also develops children's reasoning skills, which they are developing throughout the year.

Before they start this unit, it is expected that children:
• know how to multiply and divide by 2, 3, 4, 5, 8 and 10
• understand related multiplication and division facts
• know how to apply their knowledge of these facts to solve problems.

ASSESSING MASTERY

Children who have mastered this unit will be able to rapidly recall all multiplication and division facts from the 1 to 12 times-tables. They will have learnt efficient strategies and be able to use these to apply their learning in context and to find solutions to word problems.

COMMON MISCONCEPTIONS	STRENGTHENING UNDERSTANDING	GOING DEEPER
Children may think that a number multiplied or divided by 0 equals the original number.	Run intervention sessions in which children practise multiplying by 0. Show the answer visually, such as by using empty plates.	Solve some multiplication and division sentences with missing numbers.
Children may confuse multiplication and addition.	Give children two numbers to add and multiply: forming two different answers.	Repeat the Strengthening Understanding activity, but also ask children to discuss how the two processes are different.
Children may not know whether to multiply or divide when solving a problem.	Ask children to highlight key information in the word problem and draw a representation/use concrete objects to help solve it.	Children can make up their own word problems to fit a multiplication and division sentence.

Unit 5: Multiplication and division ❶

Use these pages to introduce the unit focus to children as a whole class. You can use the different characters to explore different ways of working, and to begin to discuss and develop children's reasoning skills relating to multiplication and division. Talk through the key learning points, which the characters mention, and the key vocabulary. Do children have any misconceptions? Do they understand what the vocabulary means? A classroom display showing all of the key information, particularly the times-tables, will support children throughout this unit.

STRUCTURES AND REPRESENTATIONS

Number line: The number line is an effective way to represent multiplication and division. It shows the grouping clearly and helps children practise counting on or back in groups.

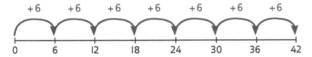

Arrays: Arrays visually show multiplication and division. They are particularly clear at showing commutativity, such as $2 \times 5 = 5 \times 2$.

KEY LANGUAGE

There is some key language that children will need to know as part of the learning in this unit.

➜ times-table, times, times by

➜ multiply (×), multiple, multiply by

➜ divide (÷), divide by

➜ grouping, groups of, lots of

➜ sharing, share, equal, equally

➜ number facts, number sentences, multiplication facts/sentences, division facts/sentences, fact family, related fact

➜ number line, array, ten frame

➜ product, factor

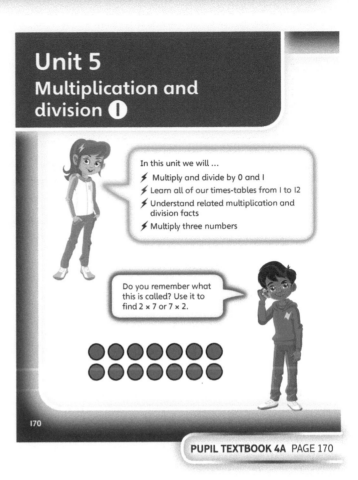

PUPIL TEXTBOOK 4A PAGE 170

PUPIL TEXTBOOK 4A PAGE 171

Multiples of 3

Learning focus

In this lesson, children will learn to name and find multiples of 3 and non-multiples of 3.

Before you teach

- Can children count in 3s?
- Can children show you how to make equal groups of 3 items (for example, counters)?
- Do children recognise the multiplication and division signs?

NATIONAL CURRICULUM LINKS

Year 4 Number – multiplication and division

Recall multiplication and division facts for multiplication tables up to 12 × 12.

ASSESSING MASTERY

Children can name and identify numbers that are multiples of 3.

COMMON MISCONCEPTIONS

Children may think that a number with a digit '3' (such as 31 or 23) is a multiple of 3. Ask:
- *When you count up in 3s, do all the numbers have a 3 digit?*

STRENGTHENING UNDERSTANDING

Provide cubes or counters and ask children, throughout, to use them to make groups of 3s. This will support the concept of multiples of 3 as 'groups of 3'.

GOING DEEPER

Challenge children to explore the relationship between '*x* groups of 3' and '3 groups of *x*'. They can use cubes or counters to demonstrate their findings.

KEY LANGUAGE

In lesson: groups of, multiple, **product**, **factor**, equal, fact

Other language to be used by teacher: multiplication, division, multiply, times, divide, fact family

STRUCTURES AND REPRESENTATIONS

Factor-factor-product triangle, number line

RESOURCES

Mandatory: cubes or counters

Optional: 100 squares

 In the eTextbook of this lesson, you will find interactive links to a selection of teaching tools.

Quick recap 🔄

As a class, count together in 3s from 0 to 30. Then play 'Can you hear my mistake?', counting in 3s and including a deliberate mistake for children to find. For example:

0, 3, 6, 9, 15, 18, 21, 24, 27, 30 or 0, 3, 6, 9, 10, 13, 15

Discover

WAYS OF WORKING Pair work

ASK

- Question ① a): *Can you show me one group of 3? Now show me another. How many groups of 3 do you think you can make? Now can you show me how to share all the counters into 3 equal groups?*
- Question ① b): *Can you write a multiplication fact about your groups of 3? Can you write a division fact to show how you can group the counters?*

IN FOCUS Questions ① a) gives children the opportunity to explore making 'x groups of 3' and '3 groups of x' in relation to multiplication and division. This demonstrates that the total number of counters is a multiple of 3.

PRACTICAL TIPS Support children in using real counters or cubes to work through the task. You could also act out the scenario in real life with 12 children being arranged first in 4 groups of 3, and then in 3 groups of 4.

ANSWERS

Question ① a): 12 is 4 groups of 3 or 3 groups of 4.

Question ① b): $4 \times 3 = 12$; $3 \times 4 = 12$; $12 \div 3 = 4$; $12 \div 4 = 3$

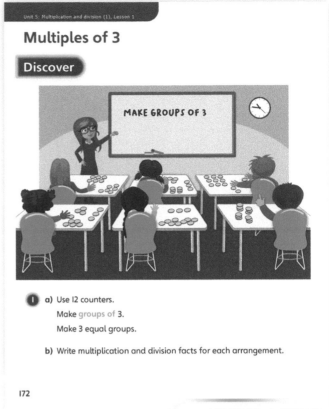

Multiples of 3

Discover

① a) Use 12 counters.
   Make groups of 3.
   Make 3 equal groups.

b) Write multiplication and division facts for each arrangement.

172

PUPIL TEXTBOOK 4A PAGE 172

Share

WAYS OF WORKING Whole class teacher led

ASK

- Question ① a): *What do you think the word 'multiple' means in the sentence '12 is a multiple of 3'?*
- Question ① a): *What is the same and what is different about each arrangement?*
- Question ① b): *How does this triangle show the relationship between 3, 4 and 12?*
- Question ① b): *In what order can you multiply the factors? In what order can you divide?*

IN FOCUS Question ① a) gives children an understanding of equal groups of counters representing factors of a given number. In question ① b), children relate fact families to the factor-factor-product relationship. Reinforce the fact that 3 is a factor, 4 is a factor and 12 is the product. Remind children of this regularly and ask them to repeat it until they are familiar with the structure.

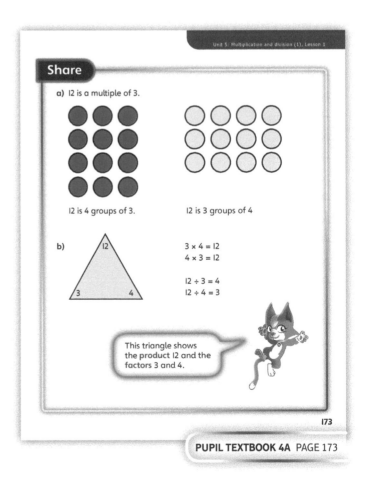

Share

a) 12 is a multiple of 3.

12 is 4 groups of 3.   12 is 3 groups of 4

b)

$3 \times 4 = 12$
$4 \times 3 = 12$

$12 \div 3 = 4$
$12 \div 4 = 3$

This triangle shows the product 12 and the factors 3 and 4.

173

PUPIL TEXTBOOK 4A PAGE 173

Think together

WAYS OF WORKING Whole class teacher led (I do, We do, You do)

ASK

- Question **1**: *What number comes next in the count?*
- Question **2**: *Do you know the factors? Do you know the product? How could you work out the missing product? Can you think of more than one way?*
- Question **3** b): *Are all the numbers a multiple of 3?*

IN FOCUS Question **1** uses the number line model to support children with counting up in steps of 3. Question **2** requires children to demonstrate an understanding of factor-factor-product families in order to find a missing product when both factors are known. In question **3** b), children apply what they have learned about equal groups in order to identify which numbers are not multiples of 3.

STRENGTHEN Rehearse the counting in 3s pattern by counting on in 3s together. Practise stopping at different multiples of 3 to build children's confidence about which numbers are in the count.

DEEPEN Challenge children to find all the multiples of 3 on a 100 square and to describe any patterns that they notice.

ASSESSMENT CHECKPOINT Question **2** assesses whether children can understand and use the factor-factor-product relationship to find a missing product and generate the related fact family.

ANSWERS

Question **1** a): 0, 3, 6, 9, 12, 15, 18, 21, 24, 27

Question **2**: 7 × 3 = 21; 3 × 7 = 21; 21 ÷ 3 = 7; 21 ÷ 7 = 3

Question **3** a): There are 20 cubes. The children can make 6 towers with 2 cubes left over; they cannot use all the cubes to make towers of 3.

Question **3** b): 13 and 23 are not multiples of 3.

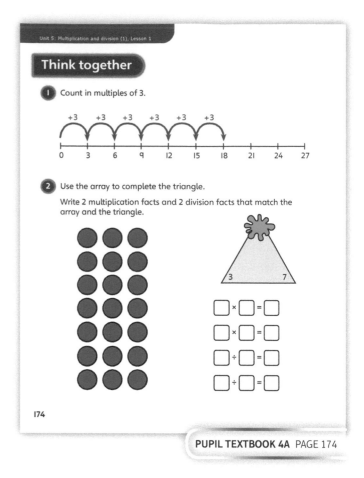

PUPIL TEXTBOOK 4A PAGE 174

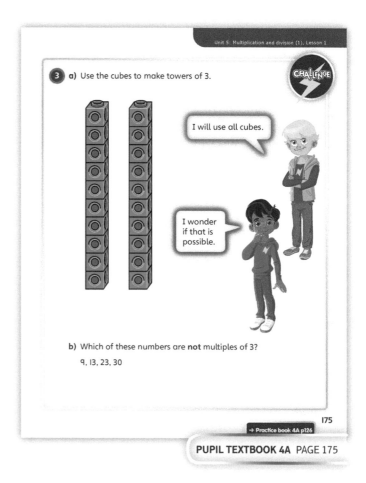

PUPIL TEXTBOOK 4A PAGE 175

Practice

WAYS OF WORKING Independent thinking

IN FOCUS Question ❶ explores the 'x groups of 3' and '3 groups of x' relationship to show multiples of 3. In question ❷, children use the factor-factor-product relationship to generate multiplication and division fact families for given arrays and in question ❸ they find the product of two given factors. Question ❹ asks children to divide by 3. They can use the factor-factor-product relationship to support them with this, drawing any representations that they think will be helpful. In question ❺, children use the number line model to complete a count in multiples of 3 and in question ❻, they use the 100 square to explore multiples of 3 beyond 36.

STRENGTHEN Provide children with 100 squares and examine the first three rows together, supporting children in identifying and rehearsing the multiples of 3.

DEEPEN Ask children to explore multiples of 3 that are greater than 200. Challenge them to describe a good way of making sure they have found every single one.

ASSESSMENT CHECKPOINT Use question ❸ to assess whether children can use given factors to generate a factor-factorproduct fact family.

ANSWERS Answers for the **Practice** part of the lesson can be found in the *Power Maths* online subscription.

Reflect

WAYS OF WORKING Independent thinking

IN FOCUS The **Reflect** part of the lesson prompts children to explore which of a series of given numbers are multiples of 3 and which are not.

ASSESSMENT CHECKPOINT Assess whether children can accurately justify their decision about which of the numbers is a multiple of 3.

ANSWERS Answers for the **Reflect** part of the lesson can be found in the *Power Maths* online subscription.

After the lesson ⏸

- Play 'I am thinking of a number'. Say: *My number is a multiple of 3 that is greater than 10*, and ask children to make sensible suggestions. Allow yes / no questions, such as '*Is it greater than x?*', '*Is one of the digits a 5?*', and so on.

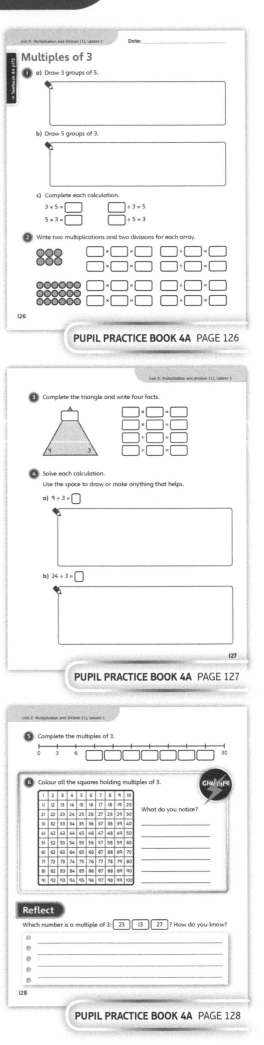

PUPIL PRACTICE BOOK 4A PAGE 126

PUPIL PRACTICE BOOK 4A PAGE 127

PUPIL PRACTICE BOOK 4A PAGE 128

Multiply and divide by 6

Learning focus

In this lesson, children will learn what it means to multiply and divide by 6. They will use a range of strategies to support their understanding.

Before you teach

- Can children count in 3s and do they know their 3 times-table?
- Can children count in 6s and do they know their 6 times-table?

NATIONAL CURRICULUM LINKS

Year 4 Number – multiplication and division

Recall multiplication and division facts for multiplication tables up to 12 × 12.

ASSESSING MASTERY

Children can confidently multiply and divide numbers by 6. They can make links, such as multiplying by 3 then doubling, or multiplying by 5 and then adding one more lot of the number they were multiplying by.

COMMON MISCONCEPTIONS

Children may not realise that 4 × 6 is equivalent to 6 × 4. Ask:
- *Can you use counters to represent each of these calculations as an array? What do you notice?*

When counting in 6s, children sometimes start by counting 0. Ask:
- *What number should we start on when counting in 6s?*

STRENGTHENING UNDERSTANDING

Children may occasionally lose count, for example, when working out 6 × 6 mentally, they may end up counting too many 6s, or too few 6s. Encourage children to use a number line and follow the jumps with their fingers. Children can also use counters in groups of 6 to see the direct link between the objects and counting on in 6s on a number line.

Knowledge of times-tables is very important in this lesson: ensure that children get daily support with this. You could also ask for some home support from parents or guardians.

GOING DEEPER

Challenge children to use multiplication facts that they know to work out other multiplication facts. For example, can they use 12 × 6 to work out 13 × 6 or 24 × 6?

KEY LANGUAGE

In lesson: multiply (×), divide (÷), how many, count on, number line, grouped, total, calculation, length, width, centimetre (cm)

Other language to be used by teacher: number sentence, times-table, equal, array

STRUCTURES AND REPRESENTATIONS

Number lines, arrays

RESOURCES

Mandatory: counters

Optional: base 10 equipment, dice

 In the eTextbook of this lesson, you will find interactive links to a selection of teaching tools.

Quick recap 🔄

Ask children to take it in turns to roll three dice. Keep going until someone rolls double 6. Agree that double 6 is 12.

Ask: *If someone rolls three 6s, what would that score be?*

Discover

Unit 5: Multiplication and division (1), Lesson 2

Multiply and divide by 6

Discover

WAYS OF WORKING Pair work

ASK

- Question ➊ a): *What multiplication sentence would match this question?*
- Question ➊ b): *Could drawing an array help us?*

IN FOCUS For question ➊ a), observe carefully which children can confidently count in 6s, and which count the eggs individually. This will tell you who will need some extra support during the lesson, and following the lesson. It is an important assessment opportunity.

PRACTICAL TIPS For question ➊ b) children could draw the boxes and eggs (in arrays). This will give them more of a visual understanding of the problem.

ANSWERS

Question ➊ a): 7 × 6 = 42. There are 42 eggs in total in the boxes.

Question ➊ b): 30 ÷ 6 = 5. So 5 egg boxes can be filled by the tray of eggs.

➊ a) Toshi has 7 full egg boxes.

How many eggs are there in total inside these boxes?

b) Under the table is a tray of 30 eggs.

How many egg boxes can be filled using the eggs in the tray?

176

PUPIL TEXTBOOK 4A PAGE 176

Share

WAYS OF WORKING Whole class teacher led

ASK

- Question ➊ a): *Why do you count in 6s and not 1s? Can counting in 2s get the answer? Why can it be seen as a multiplication?*
- Question ➊ b): *Is it easier to count on or back in 6s?*

IN FOCUS Question ➊ a) provides children with a great opportunity to practise using number lines. Ask children to put their fingers on the number line and practise counting on and back in 6s. They will see the repeated addition and link it with the problem they have been trying to find the solution for.

In question ➊ b), children can count back in 6s from 30 or count on, starting from 6. Model the counting back on the number line and also on fingers, i.e. 30, 24, 18, 12, 6 (one on each finger of a hand).

Share

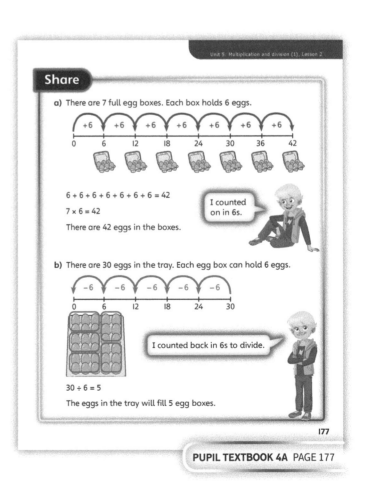

a) There are 7 full egg boxes. Each box holds 6 eggs.

6 + 6 + 6 + 6 + 6 + 6 + 6 = 42

7 × 6 = 42

There are 42 eggs in the boxes.

I counted on in 6s.

b) There are 30 eggs in the tray. Each egg box can hold 6 eggs.

30 ÷ 6 = 5

The eggs in the tray will fill 5 egg boxes.

I counted back in 6s to divide.

177

PUPIL TEXTBOOK 4A PAGE 177

211

Think together

WAYS OF WORKING Whole class teacher led (I do, We do, You do)

ASK

- Question **1**: *How did the pictures help?*
- Question **3**: *What strategies did you find worked well for you in this question?*
- Question **3**: *Can you explain your answer to a friend?*

IN FOCUS For question **3**, some children may draw a number line to help them work out the answer. This is fine, but also make sure they understand how the array can help them (an important learning point).

STRENGTHEN At each stage, use concrete objects, such as counters, alongside a number line to help children see the link with repeated addition; counting on in 6s and then reinforcing the multiplication sentence.

DEEPEN Challenge children by asking them to investigate: *How could you work out 13 × 6? How could you work out 120 ÷ 6? How could you work out 50 × 6?*

ASSESSMENT CHECKPOINT Question **2** will allow you to assess children's ability to explain what it means to multiply or divide by 6, and use their knowledge of counting in 6s to work out the answers.

ANSWERS

Question **1**: 6 + 6 + 6 + 6 + 6 + 6 + 6 + 6 = 48
8 × 6 = 48. There are 48 eggs in total.

Question **2** a): 4 × 6 = 24

Question **2** b): 36 ÷ 6 = 6

Question **3** a): 3 baskets of 18 apples.
Each basket has 18 apples arranged as a 3 × 6 array.
3 lots of 3 × 6 arrays is the same as a 9 × 6 array.
So the answer is: 9 × 6 = 54.

Question **3** b): There would be 72 apples in 4 baskets of apples.

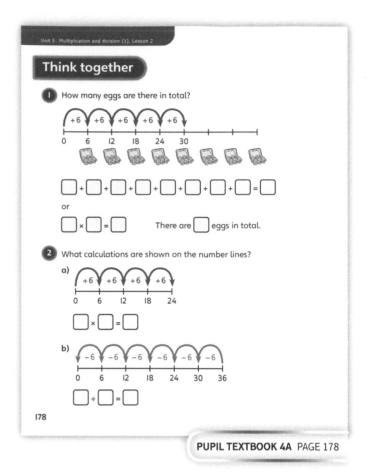

PUPIL TEXTBOOK 4A PAGE 178

PUPIL TEXTBOOK 4A PAGE 179

Practice

WAYS OF WORKING Independent thinking

IN FOCUS Question **5** is a fantastic way to put the learning in this lesson into a measurement context. It will involve higher-order thinking; discussion in pairs will help promote this.

STRENGTHEN Run some intervention exercises in which children practise counting in 6s, and also answer quick-fire questions, such as 4 × 6, 8 × 6 and so on.

If children are confident with their 3 times-table, you could teach them the method of multiplying by 3 and then doubling.

DEEPEN To deepen learning, explore how to work out 2 × 6 + 5 × 6 as one multiplication. They could show this by adding arrays together, and finding that it is the same as 7 × 6.

You could also challenge children by giving them an answer, such as 24, and asking them to think of as many questions for it as they can. They may start with simple ones, such as 4 × 6, but then may go on to think of multi-step questions, for example (8 × 6) ÷ 2.

ASSESSMENT CHECKPOINT Questions **3** and **4** provide a good opportunity to assess whether children have understood how to multiply and divide by 6. Can children represent a question as a multiplication sentence involving × 6 and can they use counting in 6s to work out the correct answer to the multiplication?

ANSWERS Answers for the **Practice** part of the lesson can be found in the *Power Maths* online subscription.

Reflect

WAYS OF WORKING Pair work

IN FOCUS This question is important because children have to use their knowledge and construct their own questions. This will involve some deeper thinking.

ASSESSMENT CHECKPOINT Check whether children have appropriate methods and if they are reasoning effectively (they may have drawn diagrams to support their reasoning).

ANSWERS Answers for the **Reflect** part of the lesson can be found in the *Power Maths* online subscription.

After the lesson ⏸

- Can children form number sentences involving × 6 or ÷ 6 from a word question?
- Do children know the method of multiplying by 3 and then doubling?
- How can children get extra practice – interventions, home support?

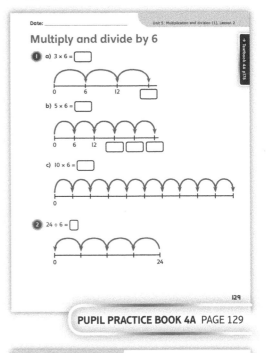

PUPIL PRACTICE BOOK 4A PAGE 129

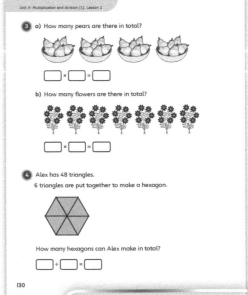

PUPIL PRACTICE BOOK 4A PAGE 130

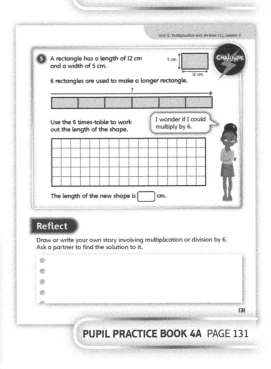

PUPIL PRACTICE BOOK 4A PAGE 131

6 times-table and division facts

Learning focus

In this lesson, children will focus on learning their 6 times-table. Children should be able to recite it and also learn the associated multiplication and division facts.

Before you teach

- How did children get on in the previous lesson?
- How will you link multiplying and dividing by 6 with the 6 times-table?
- In what ways can you encourage children to learn the 6 times-table off by heart?

NATIONAL CURRICULUM LINKS

Year 4 Number – multiplication and division

Recall multiplication and division facts for multiplication tables up to 12 × 12.

ASSESSING MASTERY

Children can demonstrate a rapid recall of multiplication and associated division facts from the 6 times-table. They can use their knowledge to find solutions and clearly explain how they worked out the answers.

COMMON MISCONCEPTIONS

To work out whether 8 × 6 is greater than or equal to 3 × 6, children sometimes think you have to work out each multiplication fact. Ask:

- *Can you reason why 8 groups of 6 has to be greater than 3 groups of 6 without working them out?* Children could draw arrays to help them understand this.

STRENGTHENING UNDERSTANDING

Multiplication and division facts must be constantly reinforced. Show children facts from the 6 times-table pictorially or by using equipment, such as counters or base 10 equipment. For example, 2 × 6 could be represented by 2 towers of 6 cubes.

Ensure that children have regular opportunities to practise rapid recall of the multiplication and division facts, but explain that they need an understanding of what they mean too.

GOING DEEPER

Challenge children to work out missing numbers in number sentences. For example, 12 ÷ 6 < ☐ × 6.

KEY LANGUAGE

In lesson: times-table, multiplication fact, count on, number line, add, divide (÷), multiply (×), total

Other language to be used by teacher: number facts, number sentence, calculation

STRUCTURES AND REPRESENTATIONS

Number lines, arrays

RESOURCES

Mandatory: counters

Optional: base 10 equipment

 In the eTextbook of this lesson, you will find interactive links to a selection of teaching tools.

Quick recap

As a class, count together on and back in multiples of 6 from 0 to 60.

Discover

6 times-table and division facts

WAYS OF WORKING Pair work

ASK

- Question ① a): *What is a good method of remembering these number facts?*
- Question ① b): *How is multiplication linked to division?*

IN FOCUS This question is important because it draws on children's knowledge of multiplying by 6 and then using this to generate the 6 times-table. Children then explore how multiplication facts are related to division facts.

PRACTICAL TIPS Show children the 6 times-table on your maths display. Cover up different parts each day: some without answers, some with answers or all of the multiplication missing but the answer given.

ANSWERS

Question ① a): Full set of 6 times-table shown.

$0 \times 6 = 0$	$7 \times 6 = 42$
$1 \times 6 = 6$	$8 \times 6 = 48$
$2 \times 6 = 12$	$9 \times 6 = 54$
$3 \times 6 = 18$	$10 \times 6 = 60$
$4 \times 6 = 24$	$11 \times 6 = 66$
$5 \times 6 = 30$	$12 \times 6 = 72$
$6 \times 6 = 36$	

Question ① b): The division fact is $42 \div 6 = 7$.

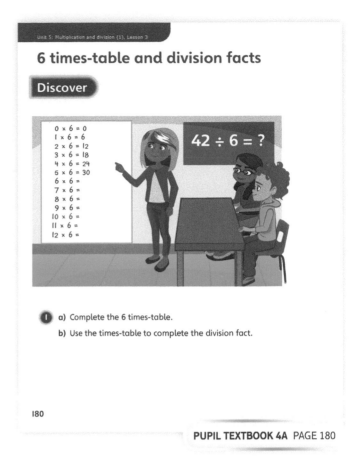

PUPIL TEXTBOOK 4A PAGE 180

Share

WAYS OF WORKING Whole class teacher led

ASK

- Question ① a): *How does the number line help you?*
- Question ① a): *Why do you need to learn your times-tables?*
- Question ① b): *Which multiplication fact from the times-table will help you? Why?*

IN FOCUS Question ① b) gives children the opportunity to explore division as the inverse of multiplication, and to generate a multiplication and division fact family in the context of the 6 times-table.

PUPIL TEXTBOOK 4A PAGE 181

Think together

Think together

WAYS OF WORKING Whole class teacher led (I do, We do, You do)

ASK

- Question ③: *How does the 5 times-table help you?*
- Question ③: *How do you know your answers in this section are correct?*
- Question ③: *Can you explain your methods?*

IN FOCUS The representation in question ① is a really important aspect of this lesson. Children, all too often, simply learn times-tables without understanding what they mean. Visual representations are key to children's understanding.

STRENGTHEN Run some intervention sessions with children who require it. Give children a fact from the 6 times-table and ask them to draw an array to represent it. Focus first on the key known facts 1×6, 2×6 and 10×6, and then move on to 5×6.

DEEPEN In question ③, discuss with children how the 5 times-table can help them work out the answers to the 6 times-table: they could multiply a number by 5 and then add on one more lot of that number. You could also challenge children to write some word problems based on the 6 times-table.

ASSESSMENT CHECKPOINT Use question ② to assess whether children can use arrays to derive multiplication and division fact families for the 6 times-table.

ANSWERS

Question ①: $10 \times 6 = 60$; $6 \times 10 = 60$
There are 60 pencils.

Question ② a): $2 \times 6 = 12$, $6 \times 2 = 12$, $12 \div 2 = 6$, $12 \div 6 = 2$

Question ② b): $4 \times 6 = 24$, $6 \times 4 = 24$, $24 \div 4 = 6$, $24 \div 6 = 4$

Question ③ a): $3 \times 5 = 15$, $3 \times 6 = 18$
The array shows 3 rows of 5 (3×5) and then 1 more in each row (3×6), or it shows 5 columns of 3 and then 1 more column of 3.

Question ③ b): You can multiply by 5 and then add on one more lot of the number you are multiplying.

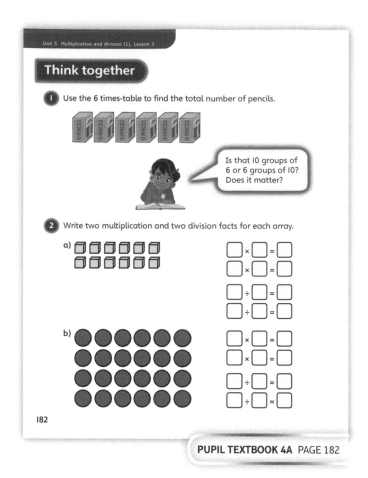

PUPIL TEXTBOOK 4A PAGE 182

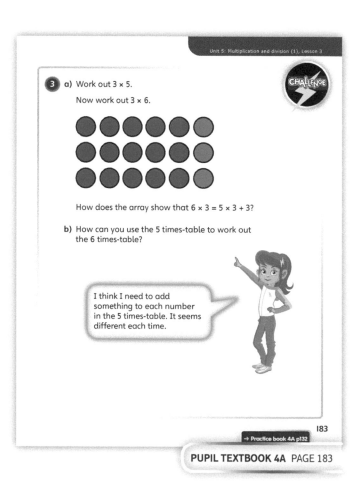

PUPIL TEXTBOOK 4A PAGE 183

Practice

WAYS OF WORKING Independent thinking

IN FOCUS Question 3 has some number sentences in which the answer appears first, such as ☐ = 6 × 10. This can often throw children. Ask them if there is another way they can write the number sentence. Showing both ☐ = 6 × 10 and 6 × 10 = ☐ is a great learning point for children.

STRENGTHEN Question 4 uses number tracks to show the 6 times-table and corresponding division facts. Provide number lines for children to count along in order to make the link with repeated addition or subtraction. Encourage children to observe how the answers to part a) can help them to find the answers for part b).

DEEPEN Challenge children to complete some 6 times-table multiplication and division facts which are littered with mistakes. Ask children to become the teacher and spot the mistakes. For example, 7 × 6 = 36, 66 ÷ 6 = 1.

As an extension, ask children to suggest why the mistakes were made (which would get them to reason out the misconceptions).

THINK DIFFERENTLY Question 5 asks children to use a multiplication fact from the 6 times-table to solve a calculation that is beyond the known times-table. Encourage them to discuss possible strategies and to explain how they found the answer.

ASSESSMENT CHECKPOINT Can children recognise facts from the 6 times-table from arrays or number lines, like for example in questions 2 and 4? Children should start to be developing a secure knowledge of their 6 times-table, both multiplication and associated division facts.

ANSWERS Answers for the **Practice** part of the lesson can be found in the *Power Maths* online subscription.

Reflect

WAYS OF WORKING Pair work

IN FOCUS Make this reflective exercise into a whole-class competition. Children really enjoy times-table races, so you could make it a weekly event. This will really encourage children to learn their times-tables.

ASSESSMENT CHECKPOINT This is the perfect opportunity to assess which children have mastered the lesson and who needs extra times-table practice.

ANSWERS Answers for the **Reflect** part of the lesson can be found in the *Power Maths* online subscription.

After the lesson

- Can children recall their division and multiplication facts from the 6 times-table?
- Can children think of some real-life problems associated with the 6 times-table?

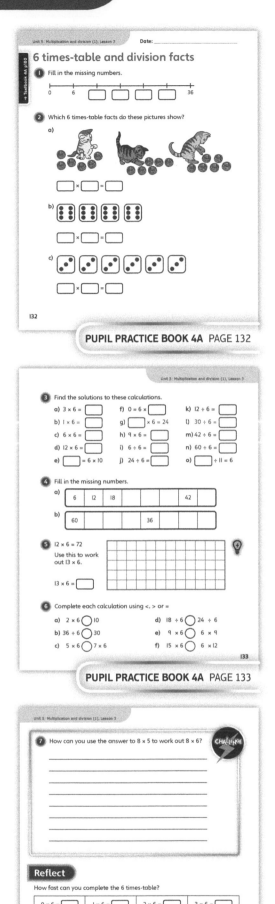

PUPIL PRACTICE BOOK 4A PAGE 132

PUPIL PRACTICE BOOK 4A PAGE 133

PUPIL PRACTICE BOOK 4A PAGE 134

Multiply and divide by 9

Learning focus

In this lesson, children will understand how they can multiply and divide a number by 9. Children will make links to the 3 and 6 times-tables.

Before you teach

- Can children count in 3s, 6s and 9s?
- Do they understand the key vocabulary?
- Can children explain what 6 × 9 means? Can they show it using a visual representation?

NATIONAL CURRICULUM LINKS

Year 4 Number – multiplication and division

Recall multiplication and division facts for multiplication tables up to 12 × 12.

ASSESSING MASTERY

Children can use a range of strategies and representations to multiply and divide numbers by 9, making links to the 3 and 6 times-tables, which will help children to work out answers quickly. Children can understand that multiplication is the inverse of division and they may use this knowledge to check answers (for example, 108 ÷ 9 = 12 can be checked by working out 12 × 9).

COMMON MISCONCEPTIONS

When multiplying and dividing by 9, children may lose track of the count. Encourage children to learn the multiplication facts and associated division facts by heart. This may involve intervention sessions and some home learning. Ask:
- *How quickly can you recall a multiplication or division? Practise them daily to increase your recall.*

STRENGTHENING UNDERSTANDING

Children often learn multiplication and division facts without understanding what they mean. Work with concrete objects to show them clearly. Use a number line to visually show repeated addition and repeated subtraction by counting on and back in 9s. Times-table practice is vital throughout this unit – run daily challenges to encourage children to memorise them.

GOING DEEPER

Linking multiplication and division facts to measures will deepen children's learning. For instance, you could set problems in which amounts of money are shared between 9 people.

KEY LANGUAGE

In lesson: multiply (×), divide (÷), cube, square, how many, number line

Other language to be used by teacher: multiplication fact/statement, division fact/statement, grouping, array

STRUCTURES AND REPRESENTATIONS

Number lines, arrays

RESOURCES

Mandatory: base 10 equipment, counters

 In the eTextbook of this lesson, you will find interactive links to a selection of teaching tools.

Quick recap 🔍

Show children the number 9 on your fingers. Then show the number 9 on your fingers in two more different ways.

Ask: *What is different and what is the same about each of these ways of showing 9?*

Discover

Unit 5: Multiplication and division (1), Lesson 4

WAYS OF WORKING Pair work

ASK

- Question ❶ a): *What facts do you know about cubes?*
- Question ❶ b): *Can you show your workings on a number line?*

IN FOCUS This question focuses on multiplying by 9 in the context of using the properties of a 3D shape. Children can then use the 9 times-table with repeated addition or multiplication to solve a related calculation.

PRACTICAL TIPS Providing children with actual puzzle cubes will really help them to visualise 6 × 9.

ANSWERS

Question ❶ a): There are 6 faces on the cube and 9 squares on each face.

Question ❶ b): $9 + 9 + 9 + 9 + 9 + 9 = 54$
$6 × 9 = 54$
There are 54 coloured squares on the whole cube.

Multiply and divide by 9

Discover

❶ a) Reena has a puzzle cube.
How many faces are on the cube?
How many squares are on each face?

b) Calculate the total number of squares on the whole cube.

184

PUPIL TEXTBOOK 4A PAGE 184

Share

WAYS OF WORKING Whole class teacher led

ASK

- Question ❶ b): *How do the number line and the arrays help you?*
- Question ❶ b): *Is multiplication quicker than repeated addition?*

IN FOCUS Discuss with children the different methods they can use in order to find solutions to question ❶ b), $9 + 9 + 9 + 9 + 9 + 9 = 54$ or $6 × 9 = 54$. This will encourage children's deep thinking and understanding about the similarities and differences between repeated addition and multiplication.

Share

a) A cube has 6 faces.
There are 9 squares on each face.

b)
0   9   18   27   36   45   54

$9 + 9 + 9 + 9 + 9 + 9 = 54$
$6 × 9 = 54$
There are 54 coloured squares in total.

I counted in 9s.

I used my 6 times-table to check.

185

PUPIL TEXTBOOK 4A PAGE 185

Think together

WAYS OF WORKING Whole class teacher led (I do, We do, You do)

ASK

- Question **2**: *How could you represent the multiplication and division visually?*
- Question **4**: *What is the biggest number you can think of that is a multiple of 9?*

IN FOCUS Question **4** looks at divisibility tests. Explain the method clearly and let children work with some examples to see how it works. Ask: *Can you explain why this method works?*

STRENGTHEN To strengthen children's understanding of question **2** a), multiply and share using counters to represent the amounts.

Another way to strengthen learning here is to model some incorrect answers and ask children to mark and correct them. For example, 5 × 9 = 43. Further to this, ask children to explain where the person who made the mistake may have gone wrong.

DEEPEN Challenge children to explore the relationship between multiplying by 9 and dividing by 9. How could they use division to work out a missing number problem such as ▢ × 9 = 72? Can they explain why? Can they make up their own problems similar to this?

You could also ask children to reason if the divisibility rule for 9 would also apply for 3.

ASSESSMENT CHECKPOINT Ask children how they know if they are doing a multiplication or division. This will show you which children understand the difference between multiplication and division and why the two processes are needed for different situations, and will show which children need more support. Ask: *What are the clues you look for?*

ANSWERS

Question **1**: 0, 9, 18, 27, 36, 45, 54, 63, 72, 81, 90

Question **2** a): Aki spends £45.

Question **2** b): Ambika can buy 3 boxes with £27.

Question **3** a): 3 × 10 = 30; 3 × 9 = 27

Question **3** b): 5 × 10 = 50; 5 × 9 = 45
7 × 10 = 70; 7 × 9 = 63

Question **3** c): The answer to the second part of each question is one lot of the number (3, 5 or 7) less than the answer to the first part of each question.

Question **4** a): 144, 279, 522

Question **4** b): Any 3-digit numbers whose digits add to 9, for example 270, 333, 153.

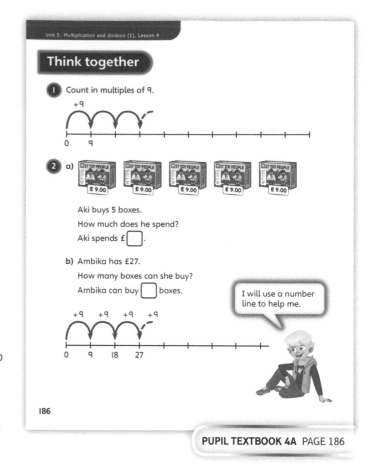

PUPIL TEXTBOOK 4A PAGE 186

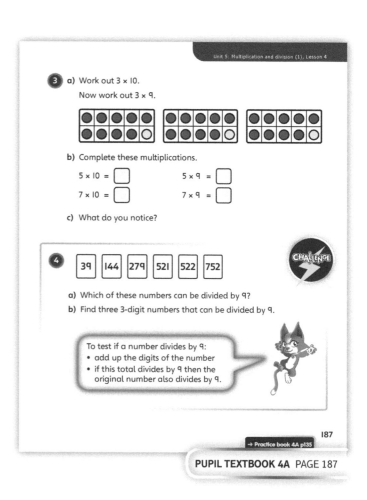

PUPIL TEXTBOOK 4A PAGE 187

Practice

WAYS OF WORKING Independent thinking

IN FOCUS Question **5** is an excellent opportunity for children to apply their knowledge of facts from the 9 times-table in the context of a story problem with money. They should recognise that they will need to carry out a division in order to find the answer.

STRENGTHEN Support children by linking the 9 times-table to other times-tables. Show them that they could work out 10 times a number and then subtract one lot of that number. This will need to be represented visually, perhaps on a number line. Also, show children that multiplying by 9 is the same as multiplying a number by 3 and then 3 again.

You may want to display the multiplication and division facts on your learning walls. This will encourage children to practise them daily.

DEEPEN Challenge children to make their own playing card problems (like in question **2**).

ASSESSMENT CHECKPOINT Assess children's ability to link the 3 and 9 times-tables by looking at answers to question **7**. Have children managed to mentally combine 3 of the towers? Or did they have to start from scratch and draw their own towers?

ANSWERS Answers for the **Practice** part of the lesson can be found in the *Power Maths* online subscription.

Reflect

WAYS OF WORKING Independent thinking

IN FOCUS This reflective exercise requires children to create a story problem to fit a division sentence. Make sure, after the question is complete, that children swap with a partner and explain their problems, reasoning why it matches.

ASSESSMENT CHECKPOINT Check if children understand whether their problem involves grouping or sharing.

ANSWERS Answers for the **Reflect** part of the lesson can be found in the *Power Maths* online subscription.

After the lesson

- Can children form multiplication statements involving × 9 from a story problem?
- Can children form division statements involving ÷ 9 from a story problem?
- Do children know at least one method to multiply and divide a number by 9?

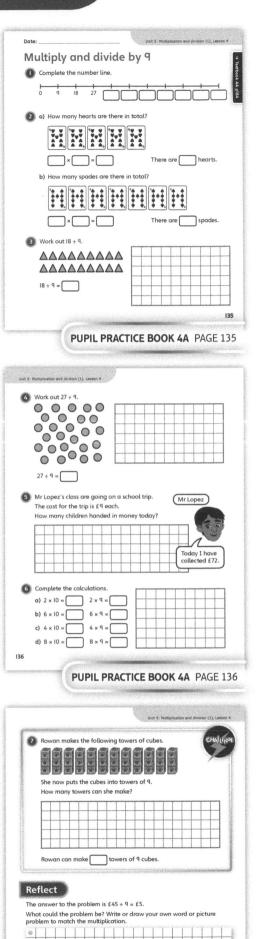

PUPIL PRACTICE BOOK 4A PAGE 135

PUPIL PRACTICE BOOK 4A PAGE 136

PUPIL PRACTICE BOOK 4A PAGE 137

9 times-table and division facts

Learning focus

In this lesson, children will focus on learning their 9 times-table. Children will also be able to recall associated division facts from the related multiplication facts.

Before you teach

- Have you got the times-tables on display in the classroom?
- How can you promote daily practice?
- Can children multiply by 9 and divide by 9?

NATIONAL CURRICULUM LINKS

Year 4 Number – multiplication and division

Recall multiplication and division facts for multiplication tables up to 12 × 12.

ASSESSING MASTERY

Children can recall multiplication and associated division facts from the 9 times-table rapidly. Children can explain how the 9 times-table links to the 10 times-table and the 3 times-table.

COMMON MISCONCEPTIONS

To work out if 6 × 9 is greater than or equal to 4 × 9, children may think you have to work out each multiplication fact separately. Ask children to reason why 6 groups of 9 has to be greater than 4 groups of 9 without working them out. Ask:
- *How can you use ten frames to show that 6 groups of 9 has to be greater than 4 groups of 9?*

STRENGTHENING UNDERSTANDING

To strengthen children's understanding of multiplication and division facts show them visually. For example, 3 × 9 could be represented with cubes (3 towers of 9). Once children have this understanding, it is important that they develop rapid recall of the multiplication and division facts.

Finally, remind children that if they are unsure of how to work out a fact from the 9 times-table they can always use their knowledge from the previous lesson. For example, to work out 3 × 9 they could do 3 × 10 = 30, then 30 – 3 = 27.

GOING DEEPER

Challenge children to apply their 9 times-table knowledge to larger numbers. For instance, you could ask them if they think 945 is a multiple of 9. They should be able to reason that (100 × 9) + (5 × 9) = 945, so it is a multiple of 9. Alternatively, they could use the divisibility rule: 9 + 4 + 5 = 18, and 18 is divisible by 9, so 945 must be divisible by 9.

KEY LANGUAGE

In lesson: times-table, facts, ten frames, multiply (×), less, greater, multiplication facts, division facts, number line, array, mathematical statements

Other language to be used by teacher: divide (÷), grouping, multiplication statement, division statement, recall, greater, number sentences

STRUCTURES AND REPRESENTATIONS

Number lines, arrays, ten frames

RESOURCES

Mandatory: counters

Optional: base 10 equipment

 In the eTextbook of this lesson, you will find interactive links to a selection of teaching tools.

Quick recap

As a class, count together on and back in multiples of 9 from 0 to 90.

Discover

WAYS OF WORKING Pair work

ASK

- Question ① a): *What can we see in the pictures? What is a good way to remember the 9 times-table?*
- Question ① b): *What calculation is needed here?*

IN FOCUS This question requires children to draw on their knowledge of multiplication and division facts from the 9 times-table to solve story problems.

PRACTICAL TIPS Show children the 9 times-table on your maths display. Cover up different parts each day: some without answers, some with answers or all of the multiplication missing but the answer given.

ANSWERS

Question ① a): $4 \times 9 = 36$
They have had 36 turns in total.

Question ① b): $63 \div 9 = 7$
They have played 7 games.

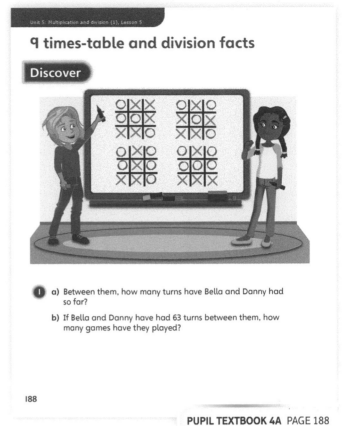

9 times-table and division facts

Discover

① a) Between them, how many turns have Bella and Danny had so far?

b) If Bella and Danny have had 63 turns between them, how many games have they played?

188

PUPIL TEXTBOOK 4A PAGE 188

Share

WAYS OF WORKING Whole class teacher led

ASK

- Question ① a): *How are the games of noughts and crosses like multiplication arrays?*
- Question ① a): *How is the multiplication fact represented with arrays?*
- Question ① b): *What multiplication fact could help you here? If you know that fact, what other facts do you know?*

IN FOCUS Undertake a whole class discussion about the methods that children have used. Explore how fact families from the 9 times-table have helped them.

Listen carefully to children's reasoning. Encourage the use of correct mathematical vocabulary.

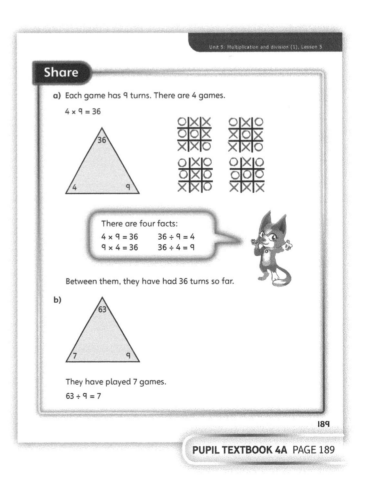

Share

a) Each game has 9 turns. There are 4 games.

$4 \times 9 = 36$

There are four facts:
$4 \times 9 = 36$ $36 \div 9 = 4$
$9 \times 4 = 36$ $36 \div 4 = 9$

Between them, they have had 36 turns so far.

b)

They have played 7 games.
$63 \div 9 = 7$

189

PUPIL TEXTBOOK 4A PAGE 189

Think together

Whole class teacher led (I do, We do, You do)

ASK

- Question **1**: *How do you know that 5 × 9 cannot be greater than 50?*
- Question **3**: *What method could you use to work out the hidden numbers? Do you find Sparks's method useful or do you prefer to use a different method?*
- Question **3**: *Explain your method clearly.*

IN FOCUS Question **3** allows children to practise methods for learning facts from the 9 times-table. Discuss Sparks's method of multiplying by 10 and then subtracting one lot. You could also suggest the method of multiplying by 3 and then by 3 again.

STRENGTHEN You may want to run some intervention sessions, practising methods children could use to work out facts from the 9 times-table. Try repeating question **3**, with different multiplication and division facts from the 9 times-table.

DEEPEN Challenge children by asking if they know their 18 times-table. For instance, they will come to realise that 18 × 3 is equivalent to 9 × 6.

ASSESSMENT CHECKPOINT Question **2** will show whether children have learnt their 9 times-table. You could also run mini-tests with children to see if they can quickly recall facts from the 9 times-table.

ANSWERS

Question **1**: $5 \times 9 = 45$ and $9 \times 5 = 45$
$45 \div 9 = 5$ and $45 \div 5 = 9$

Question **2**:
a) $0 \times 9 = 0$
b) $11 \times 9 = 99$
c) $63 \div 9 = 7$
d) $1 \times 9 = 9$
e) $12 \times 9 = 108$
f) $54 \div 9 = 6$
g) $5 \times 9 = 45$
h) $36 \div 9 = 4$
i) $8 \times 9 = 72$
j) $18 \div 9 = 2$

Question **3** a): $6 \times 9 = 54$; $7 \times 9 = 63$; $8 \times 9 = 72$

Question **3** b):
$1 \times 10 = 10$
$10 - 1 = 9$
$2 \times 10 = 20$
$20 - 2 = 18$
$3 \times 10 = 30$
$30 - 3 = 27$
$4 \times 10 = 40$
$40 - 4 = 36$
$5 \times 10 = 50$
$50 - 5 = 45$
$6 \times 10 = 60$
$60 - 6 = 54$

$7 \times 10 = 70$
$70 - 7 = 63$
$8 \times 10 = 80$
$80 - 8 = 72$
$9 \times 10 = 90$
$90 - 9 = 81$
$10 \times 10 = 100$
$100 - 10 = 90$
$11 \times 10 = 110$
$110 - 11 = 99$
$12 \times 10 = 120$
$120 - 12 = 108$

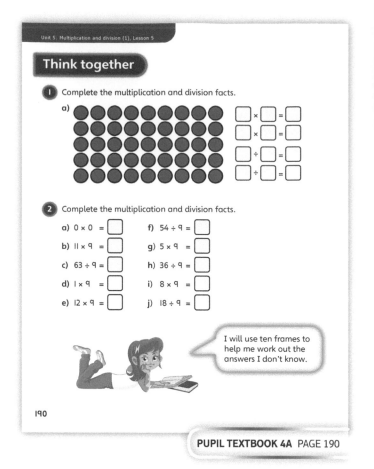

PUPIL TEXTBOOK 4A PAGE 190

PUPIL TEXTBOOK 4A PAGE 191

Practice

WAYS OF WORKING Independent thinking

IN FOCUS Question **4** requires children to draw on their knowledge of the 9 times-table and related division facts in order to complete missing number problems.

STRENGTHEN Arrays are a key representation in this lesson; suggest to children that they draw them if they are unsure of a problem. This will strengthen their learning. Also, continue to encourage rapid recall of multiplication facts via daily practice of times-tables.

DEEPEN Challenge children to think in a different way by giving them a multiple of 9, such as 27. Ask children if they can think of all the factors of 27. This will help them make more links between times-tables.

ASSESSMENT CHECKPOINT When children are completing questions **1**, **2** and **3** it is a good idea to assess children's understanding by asking them: *Tell me what you know.* Reasoning is a key step to mastery of the 9 times-table. Furthermore, look for whether children can recognise multiplication and division facts from the 9 times-table from pictures.

ANSWERS Answers for the **Practice** part of the lesson can be found in the *Power Maths* online subscription.

Reflect

WAYS OF WORKING Pair work

IN FOCUS This **Reflect** activity brings together their work on the 9 times-table. Peer checking of the answers is a great way of consolidating this lesson.

ASSESSMENT CHECKPOINT Check whether children have instant recall of their 9 times-table. You could do a thumbs assessment (thumbs up, down or in the middle) at the end of the lesson, allowing children to self-assess how they got on in the activity. Furthermore, look carefully at children's written facts. Are they accurate? Are there any misconceptions?

ANSWERS Answers for the **Reflect** part of the lesson can be found in the *Power Maths* online subscription.

After the lesson ⏸

- Can children instantly recall their division and multiplication facts from the 9 times-table?
- Which facts are children struggling with? How can these be reinforced further?
- Do children know the connection between the 3, 9 and 10 times-tables?

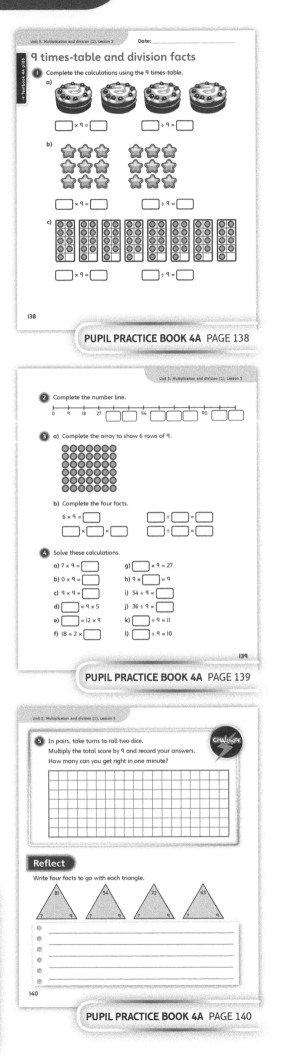

PUPIL PRACTICE BOOK 4A PAGE 138

PUPIL PRACTICE BOOK 4A PAGE 139

PUPIL PRACTICE BOOK 4A PAGE 140

The 3, 6 and 9 times-tables

Learning focus

In this lesson, children will explore the relationship between multiples of 3, multiples of 6 and multiples of 9, and develop strategies to improve their own times-tables knowledge.

Before you teach

- Can children name a multiple of 3?
- Can they find more than two multiples of 6?
- Can they list multiples of 9?

NATIONAL CURRICULUM LINKS

Year 4 Number – multiplication and division

Recall multiplication and division facts for multiplication tables up to 12 × 12.

ASSESSING MASTERY

Children can explain how multiples of 3, multiples of 6 and multiples of 9 are related to each other.

COMMON MISCONCEPTIONS

Children may think that multiples of 6 cannot also be multiples of 3 or 9. Ask:
- *I am going to say the 3 times-table. Can you hear any numbers that are also in the 6 times-table or the 9 times-table?*

STRENGTHENING UNDERSTANDING

Use **Think together** question ❸ as the basis of a strategy for supporting children in developing an improved knowledge of multiplication facts.

GOING DEEPER

Challenge children to fully explore all the patterns they can find in each times-table on the 100 square.

KEY LANGUAGE

In lesson: multiple, odd, even, pattern,

Other language to be used by teacher: sort, times-table, multiplication fact

STRUCTURES AND REPRESENTATIONS

Sorting circles, 100 square

RESOURCES

Mandatory: number cards or sticky notes, 100 squares, counters of 3 different colours

 In the eTextbook of this lesson, you will find interactive links to a selection of teaching tools.

Quick recap 🔎

As a class, count together on and back in 3s from 0 to 30. Then count on and back in 6s from 0 to 60. Finally, count on and back in 9s from 0 to 90.

Discover

WAYS OF WORKING Pair work

ASK

- Question ❶: *What do you notice about the sorting circles? Can you see the different sections?*
- Question ❶ b): *What do you think the red part on the left represents? What about the part in the middle?*

IN FOCUS Question ❶ uses sorting circles as a tool for organising and classifying numbers, including a multiple of 3 and a multiple of both 3 and 6. It also includes a number which is not a multiple of 3 or 6.

PRACTICAL TIPS Work through the activity in the classroom using sorting circles and numbers on number cards or sticky notes so that children can move each number around as they discuss their thinking and decisions.

ANSWERS

Question ❶ a): 15 is a multiple of 3, but not of 6.

12 is a multiple of both 3 and 6.

10 is not a multiple of 3 or 6.

Question ❶ b): The large section of the blue circle remains empty.

All multiples of 6 are also multiples of 3. See image from **Share** to illustrate this.

PUPIL TEXTBOOK 4A PAGE 192

Share

WAYS OF WORKING Whole class teacher led

ASK

- Question ❶ a): *How does the diagram show that 12 is a multiple of both 3 and 6? Why is 15 not in the middle section?*
- Question ❶ a): *Why is 10 outside both circles?*
- Question ❶ b): *What do you notice about the blue section on the right?*
- Question ❶ b): *Can you find numbers from the 6 times-table in the 3 times-table?*

IN FOCUS In question ❶ a), children identify which number is a multiple of 3 and 6, which number is a multiple of 3 only, and which number is a multiple of neither 3 nor 6. In question ❶ b), children explore how the 3 and 6 times-tables are related. They should find that every multiple of 6 is also a multiple of 3.

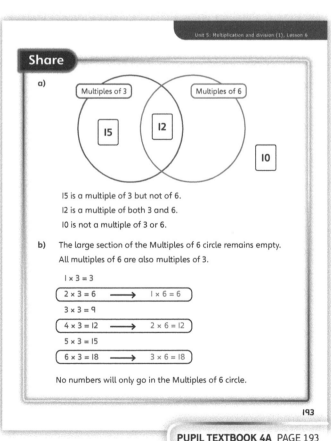

PUPIL TEXTBOOK 4A PAGE 193

Think together

WAYS OF WORKING Whole class teacher led (I do, We do, You do)

ASK

- Question **1**: *What multiples of 3 do you know? What multiples of 6? Can you cover them on the 100 square? What patterns do you notice?*
- Question **2**: *What numbers could you try this out with? How could the 100 square in question 1 help? Have you tried listing the 3, 6 and 9 times-tables?*
- Question **3**: *What facts do you know well? What facts do you want to learn next?*

IN FOCUS Question **1** uses the 100 square to support children in exploring the link between the 3, 6 and 9 times-tables. In question **2**, they investigate the properties of multiples of 3, 6 and 9 in more detail by exploring mathematical statements, some of which are not true. In question **3**, children work on strategies to improve their knowledge of multiplication and division facts.

STRENGTHEN Work through question **3** together, encouraging children to discuss how they will develop personal strategies for learning any facts they are uncertain about, in an achievable and supportive way.

DEEPEN Ask children to explore different strategies for using the 3 times-table to find a fact in the 6 or 9 times-table. Ask, for example: *If you know 4 × 3, can you use that to work out 4 × 6 or 4 × 9?*

ASSESSMENT CHECKPOINT Question **3** will allow you to assess whether children can describe and explain their own learning strategies and strengths when working with times-tables.

ANSWERS

Question **1**:

I	2	3	4	5	6	7	8	9	10
II	12	13	14	15	16	17	18	19	20
21	22	23	24	25	26	27	28	29	30
31	32	33	34	35	36	37	38	39	40
41	42	43	44	45	46	47	48	49	50
51	52	53	54	55	56	57	58	59	60
61	62	63	64	65	66	67	68	69	70
71	72	73	74	75	76	77	78	79	80
81	82	83	84	85	86	87	88	89	90
91	92	93	94	95	96	97	98	99	100

Multiples of 3 are shaded. All the multiples of 6 are also multiples of 3. All the multiples of 9 are also multiples of 3. The even multiples of 9 are also multiples of 6.

Question **2**: All statements are sometimes true.

Question **3** a): A multiple of 3 is any square where at least one of the digits is 3, 6 or 9. A multiple of 6 is a square where one of the digits is 6. A multiple of 9 is a square where one of the digits is 9.

Question **3** b): Answers will vary between children.

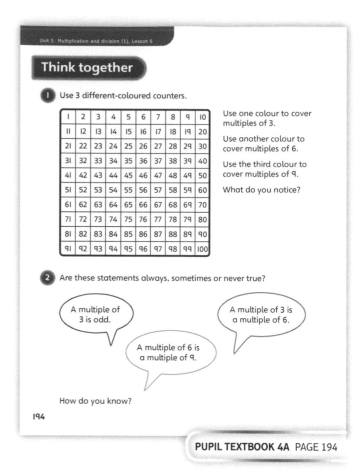

PUPIL TEXTBOOK 4A PAGE 194

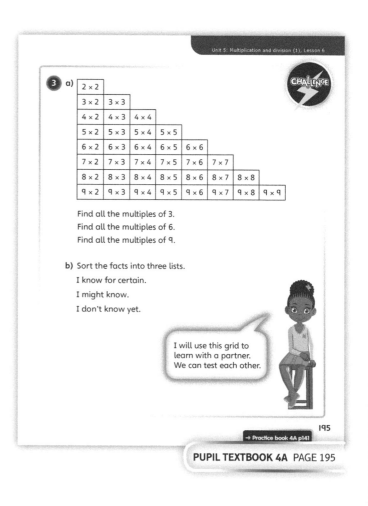

PUPIL TEXTBOOK 4A PAGE 195

Practice

WAYS OF WORKING Pair work

IN FOCUS In question ❶, children count in multiples of 3, 6 and 9. In question ❷, they use the sorting circles model to sort multiples of 3 and 9, identifying which numbers are multiples of both 3 and 9. Question ❸ requires children to think of numbers which are multiples of 3 or 9 or both. In question ❹, children explore statements about properties of multiples, some of which are not true. Question ❺ gives children the opportunity to explore equations as 'balance' statements, where the total of one side must be the same as the total of the other. They use what they have learned about the relationship between multiples of 3, 6 and 9 to find the missing numbers, ensuring each equation is balanced.

STRENGTHEN Refer children back to the number lines in question ❶, showing multiples of 3, 6 and 9. They can use these to support their thinking in later questions.

DEEPEN Question ❹ is a good opportunity for children to explore properties of multiples of 3, 6 and 9 in greater depth. Challenge them to write similar statements, which may not be true, for a partner to investigate.

ASSESSMENT CHECKPOINT Assess whether children can explain how they know whether a number is a multiple of 3, 6, 9 or none of those.

ANSWERS Answers for the **Practice** part of the lesson can be found in the *Power Maths* online subscription.

Reflect

WAYS OF WORKING Pair work

IN FOCUS The **Reflect** part of the lesson prompts children to think about their own learning and effectively develop learning strategies that they can use in future.

ASSESSMENT CHECKPOINT Assess whether children can explain their own strengths and describe challenges for their learning of times-tables.

ANSWERS Answers for the **Reflect** part of the lesson can be found in the *Power Maths* online subscription.

After the lesson

- What facts did children find tricky and what strategies have they suggested for learning them in future? How could you work on these as a class?
- Play 'I am thinking of a number'. Say: *My number is a multiple of 9 and also a multiple of 6. What could it be?* Encourage children to use what they have learned to make sensible guesses.

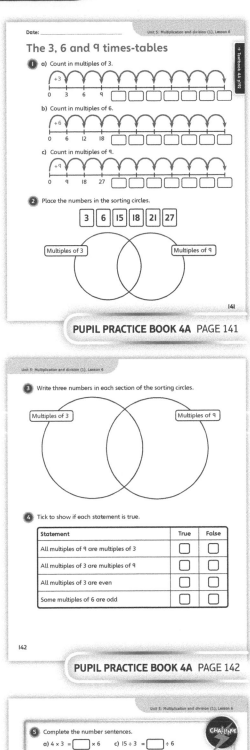

PUPIL PRACTICE BOOK 4A PAGE 141

PUPIL PRACTICE BOOK 4A PAGE 142

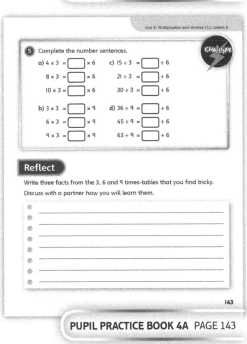

PUPIL PRACTICE BOOK 4A PAGE 143

Multiply and divide by 7

Learning focus

In this lesson, children will learn what it means to multiply and divide by 7. They will apply their knowledge to finding solutions involving real-life contexts.

Before you teach

- Have you got resources ready to support children?
- How did children get on in the last lesson?
- Which children do you think will need support in this lesson?

NATIONAL CURRICULUM LINKS

Year 4 Number – multiplication and division

Recall multiplication and division facts for multiplication tables up to 12 × 12.

ASSESSING MASTERY

Children can rapidly multiply and divide numbers by 7, as well as form multiplication and division sentences. Children can fully understand links between the 7 times-table and other times-tables, such as 5 × 7 = 35 and (5 × 5) + (5 × 2) = 35.

COMMON MISCONCEPTIONS

Children may make mistakes when counting on or back in 7s. Ask: *How do you count on or back?* Recap this.

Children may get confused with word problems – they may not understand whether they need to multiply or divide. Ask:
- *How can you use a representation to help you understand?*

Children may miscount when counting in 7s. Ask:
- *Is it helpful to count up in 7s on a number line?*

STRENGTHENING UNDERSTANDING

To strengthen children's knowledge of the 7 times-table, run daily practice of counting in 7s. Try to get support from home.

Ensure that children know the direct link between repeated addition and multiplication (use visual representations, such as the number line).

Another way to strengthen understanding would be to show children multiplication and division sentences and ask them to represent them using counters (model how to put them into an array), for example, 2 × 7 = 14 is 2 rows of 7 counters or 7 rows of 2 counters.

GOING DEEPER

Challenge children to use multiplication and division facts they know to work out more complicated ones, such as 13 × 7 or 140 ÷ 7. Make sure children explain how they worked out their answers.

KEY LANGUAGE

In lesson: multiply (×), divide (÷), how many, groups, array, number track, ten frames

Other language to be used by teacher: number sentence, times-table, equal groups, *x* groups of *y*, count in 7s

STRUCTURES AND REPRESENTATIONS

Number lines, arrays

RESOURCES

Mandatory: base 10 equipment, counters

 In the eTextbook of this lesson, you will find interactive links to a selection of teaching tools.

Quick recap 🔁

Ask children to show the number 7 on their fingers. Then ask them to show the number 7 on their fingers in two more different ways. Ask children to work with a partner and to both show 7 on their fingers. Ask: *How many fingers are you showing altogether?*

Discover

WAYS OF WORKING Pair work

ASK

- Question **1** a): *What do you notice about the groups of circles in the painting? How did you know this was a multiplication? Do you have any strategies to find the answers?*
- Question **1** b): *How did you know this was a division? Do you have any strategies to find the answers?*

IN FOCUS You may want to draw out the difference between multiplying and dividing by 7 by comparing questions **1** a) and **1** b). Ask children how they know question **1** a) is a multiplication. Ask them how they know question **1** b) is a division.

PRACTICAL TIPS For this activity, if children are counting the individual circles ask them if they can think of a more efficient method.

ANSWERS

Question **1** a): The painting is called 7s because the circles are in groups of 7s. There are also 7 groups of 7 circles.
There are 35 dotted circles.
There are 14 black circles.
There are 49 circles in total.

Question **1** b): $28 \div 7 = 4$
There are 4 groups of 7 circles.

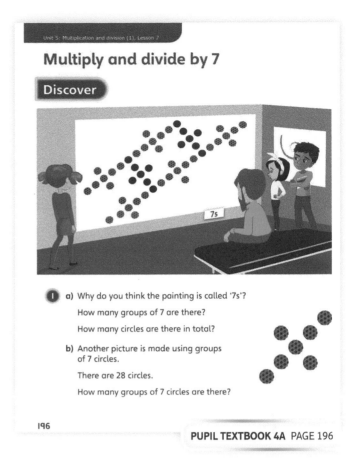

PUPIL TEXTBOOK 4A PAGE 196

Share

WAYS OF WORKING Whole class teacher led

ASK

- Question **1** a): *What method is more efficient than to count in groups of 7?*
- Question **1** b): *How does the array help you find the solution?*

IN FOCUS In question **1** b), encourage children to focus on the array. Discuss how it helps us understand the multiplication. It might be useful for you to model the multiplication on a number line and compare how they show multiplications and divisions in slightly different ways.

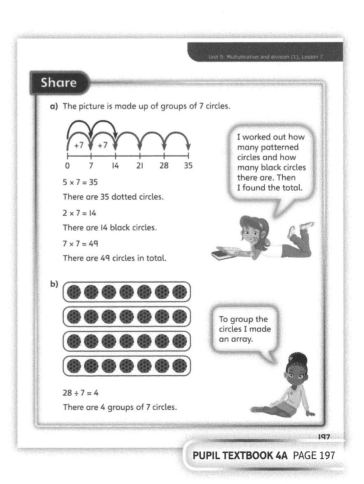

PUPIL TEXTBOOK 4A PAGE 197

Think together

WAYS OF WORKING Whole class teacher led (I do, We do, You do)

ASK

- Question **2**: *How much does one hat cost? What calculation do you need to do, a multiplication or a division?*
- Question **3**: *Will you use multiplication or division to help Lexi and Mr Jones?*

IN FOCUS Question **1** uses the number line model to support children with counting on in 7s. Children may find it helpful to refer back to this model when answering questions **2** and **3**, which require them to apply 7 times-table multiplication and division facts to word problems in the context of money and time.

STRENGTHEN For each question, have concrete objects ready for children to group and share, and to help with counting on.

Children who struggle to remember the count sequence for 7s may need a multiplication square.

DEEPEN Challenge children to find the odd one out: 714, 7, 77, 97, 105, 140.

ASSESSMENT CHECKPOINT Use question **3** to assess whether children can multiply and divide by 7 when in a real-life context.

ANSWERS

Question **1**: 0, 7, 14, 21, 28, 35, 42, 49, 56, 63, 70

Question **2** a): 6 × 7 = 42
Alex spends £42.

Question **2** b): 56 ÷ 7 = 8
Zac can buy 8 hats.

Question **3**: Lexi: 6 × 7 = 42.
There are 42 days in 6 weeks.
6 weeks is longer than 40 days.
Mr Jones: 4 × 7 = 28.
There are 7 days in a week. In 4 weeks there are 28 days.
There are 30 days in September so there are more than 4 weeks in September.

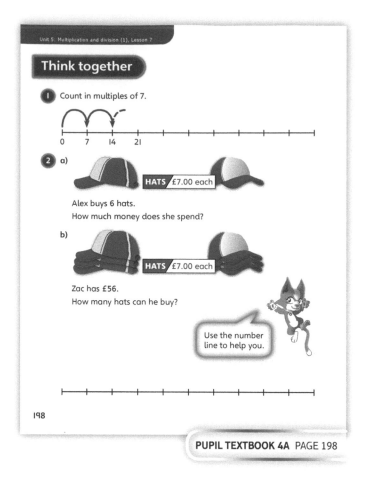

PUPIL TEXTBOOK 4A PAGE 198

PUPIL TEXTBOOK 4A PAGE 199

Practice

WAYS OF WORKING Independent thinking

IN FOCUS Question **4** is a great way to reinforce good practice with numbers. Show children counters placed messily. Ask children to count them. Then show counters placed in neat rows on ten frames. Ask children to count them again. What do they learn from this exercise?

STRENGTHEN Strengthen learning with question **2** by running intervention sessions, in which this question is repeated but with different numbers missing. After the task is complete, practise counting out loud – on and back – on the number track.

DEEPEN Challenge children to explain why a multiple of 70 is always a multiple of 7.

You could also challenge children to answer whether the following statement is correct: 'I do not need to use a written method to know that 7,280 is divisible by 7.'

Alternatively, you may want to provide further problems similar to question **6**.

THINK DIFFERENTLY Question **5** requires children to carry out a division calculation with a dividend that is not in the 7 times-table and so will have some left over.

ASSESSMENT CHECKPOINT Question **6** will allow you to assess who has mastered multiplying and dividing by 7. Children who have demonstrated excellent problem-solving skills will work with what they know and complete the steps needed to find the solution.

ANSWERS Answers for the **Practice** part of the lesson can be found in the *Power Maths* online subscription.

Reflect

WAYS OF WORKING Pair work

IN FOCUS Children show an understanding of how the numbers in the multiplication 5 × 7 = 35 can be applied to the structure of a word problem. Children could be given different 7 times-table multiplication facts as a further challenge.

ASSESSMENT CHECKPOINT Showing that they understand how to arrange the numbers in their story problem to represent 5 × 7 = 35 will demonstrate that children understand the 7 times-table and can use it to multiply and divide by 7.

ANSWERS Answers for the **Reflect** part of the lesson can be found in the *Power Maths* online subscription.

After the lesson

- How will you support children who said they found the learning difficult in this lesson?
- What strategies worked well?
- Are you continuing to update your times-table display to promote daily revision?

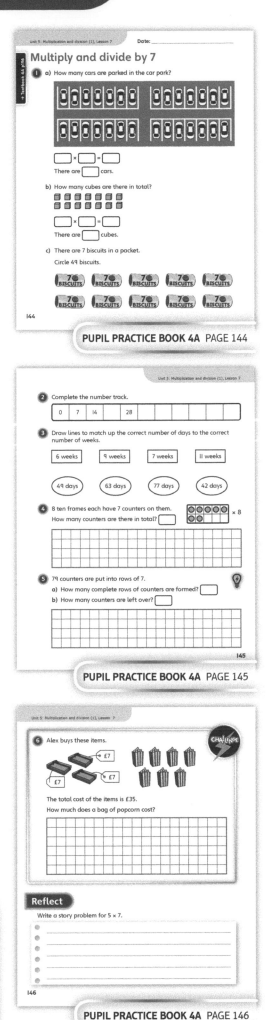

PUPIL PRACTICE BOOK 4A PAGE 144

PUPIL PRACTICE BOOK 4A PAGE 145

PUPIL PRACTICE BOOK 4A PAGE 146

7 times-table and division facts

Learning focus

In this lesson, children will focus on learning their 7 times-table. Children should be able to recite it, and learn the associated multiplication and division facts.

Before you teach

- How will you link the previous lesson to this one?
- Is your learning wall up to date?
- Could you include some visual representations (arrays or number lines) on your learning wall?

NATIONAL CURRICULUM LINKS

Year 4 Number – multiplication and division

Recall multiplication and division facts for multiplication tables up to 12 × 12.

ASSESSING MASTERY

Children can demonstrate a rapid recall of multiplication and associated division facts from the 7 times-table. Children can use their knowledge to find solutions and clearly explain how they got to the answers.

COMMON MISCONCEPTIONS

Children may make mistakes with multiplication facts, such as 7 × 6 = 43. Ask:
- *What would be a good way to revise your times-tables?*

To work out if 8 × 7 is greater than or equal to 3 × 7, children may think you have to work out each multiplication fact. Ask children to reason why 8 groups of 7 has to be greater than 3 groups of 7 without calculating. Ask:
- *How could you draw an array to help you?*

Children may also need to be reminded (link back to earlier lessons in this unit) that 1 × 7 = 7 and 0 × 7 = 0. Ask:
- *What do you recall about the learning from earlier in this unit?*

STRENGTHENING UNDERSTANDING

Multiplication and division facts must be constantly reinforced. Try and reach out to families to support children with home practice. Another good idea is to get older children to come to practise with the class (similar to paired reading but with times-tables).

Provide children with facts from the 7 times-table and ask children to represent them visually (arrays or number lines work well).

GOING DEEPER

Challenge children to create some word problems based on the 7 times-table.

KEY LANGUAGE

In lesson: times-table, multiply (×), how many, groups, array, in total, related facts, solution, solving, multiplication facts, division facts, number line, multiplication wheel

Other language to be used by teacher: number facts, number sentences

STRUCTURES AND REPRESENTATIONS

Number lines, arrays, multiplication wheel

RESOURCES

Mandatory: base 10 equipment, counters

 In the eTextbook of this lesson, you will find interactive links to a selection of teaching tools.

Quick recap ↻

As a class, count together on and back in 7s from 0 to 70.

Discover

ASK

In lesson 10, there is a whole lesson dedicated to understanding the concept of multiplying by 0 more deeply. In this lesson you can look at 0 × 7 and 1 × 7 as being part of the 7 times-table. Ask:

- Question ① a): *What do you think happens when you multiply by 0?*
- Question ① b): *How will the 7 times-table help you?*

IN FOCUS Children should recognise that the numbers on the chests look familiar because they are all in the 7 times-table. Discuss how they can use multiplying and dividing by 7 to explore questions ① a) and b).

PRACTICAL TIPS Show children the 7 times-table on your maths display. Cover up different parts each day: some without answers, some with answers or all of the multiplication missing but the answer given.

Also, you could ask children to create a visual display for the 7 times-table from arrays.

ANSWERS

Question ① a): 0 × 7 = 0
The 0 × 7 key opens the 0 chest.
8 × 7 = 56
The 8 × 7 key opens the 56 chest.
4 × 7 = 28
The 4 × 7 key opens the 28 chest.

Question ① b): Other keys needed are:
11 × 7; 1 × 7; 6 × 7; 2 × 7; 10 × 7 and 9 × 7

Share

ASK

- Question ① a): *What multiplied by 7 equals 56?*
- Question ① b): *Will you use multiplication or division to find each answer?*

IN FOCUS For question ① a) explore with children which other times-tables the numbers on the relevant chests occur in and how this could help solve the multiplications.

PUPIL TEXTBOOK 4A PAGE 200

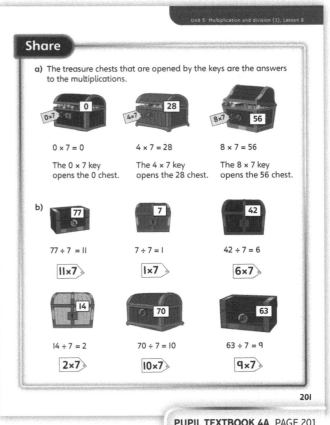

PUPIL TEXTBOOK 4A PAGE 201

Think together

WAYS OF WORKING Whole class teacher led (I do, We do, You do)

ASK

- Question ❸: *How do the 2 and 5 times-tables help us with the 7 times-table?*
- Question ❸: *How might arrays be helpful here?*

IN FOCUS Question ❷ presents a good opportunity to emphasise commutativity. Explain to children that 5 × 7 = 35 and 7 × 5 = 35, so 5 × 7 = 7 × 5. Challenge them to show you the difference using a visual representation, such as by drawing two different arrays.

STRENGTHEN Question ❷ is also an excellent opportunity to reinforce children's knowledge of fact families. Repeat this exercise when you get a chance with children who need the reinforcement. Change the starting facts and challenge them to complete the fact family.

DEEPEN Challenge children to create some word problems based on the 7 times-table.

ASSESSMENT CHECKPOINT Question ❷ will allow you to assess whether children can link multiplication facts for the 7 times-table with related division facts. Look for the correct fact families being found – children should be able to explain them to you.

ANSWERS

Question ❶: 6 × 7 = 42, 7 × 6 = 42

Question ❷ a): 5 × 7 = 35, 7 × 5 = 35, 35 ÷ 7 = 5, 35 ÷ 5 = 7

Question ❷ b): 12 × 7 = 84, 7 × 12 = 84, 84 ÷ 7 = 12, 84 ÷ 12 = 7

Question ❸ a): 5 × 3 = 15, 2 × 3 = 6, 7 × 3 = 21

Question ❸ b): 1 × 7 = 7
2 × 7 = 14
3 × 7 = 21
4 × 7 = 28
5 × 7 = 35
6 × 7 = 42
7 × 7 = 49
8 × 7 = 56
9 × 7 = 63
10 × 7 = 70
11 × 7 = 77
12 × 7 = 84

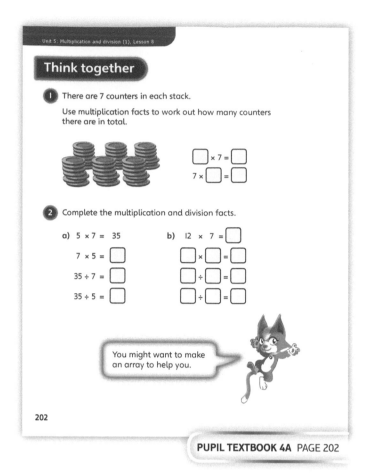

PUPIL TEXTBOOK 4A PAGE 202

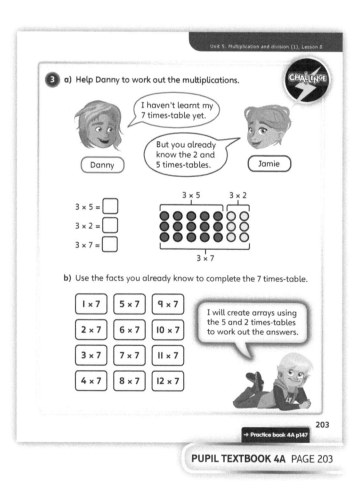

PUPIL TEXTBOOK 4A PAGE 203

Practice

WAYS OF WORKING Independent thinking

IN FOCUS In question ③, children work through a series of facts from the 7 times-table expressed as multiplication and division calculations with missing numbers. In lesson 10, there is a whole lesson dedicated to understanding the concept of multiplying by 1 and by 0. In this lesson, focus on helping students remember that $7 \times 0 = 0$ is one of the multiplication facts from the 7 times-table.

STRENGTHEN Throughout this lesson, a key learning point has been matching facts from the 7 times-table to representations or pictures. It is really important that children can do this. Provide more opportunities to practise for children who need it.

DEEPEN Challenge children by giving them some 'would you rather' statements. For example, would you rather have $6 \times £7$ or $(6 \times £3) + (6 \times £4)$?

THINK DIFFERENTLY For question ⑤, children need to solve multiplications using a multiplication wheel. This different representation may initially confuse some children, so make sure they have understood how it works.

ASSESSMENT CHECKPOINT Question ⑥ will allow you to assess which children have achieved mastery. Look for quick recall of the 7 times-table.

ANSWERS Answers for the **Practice** part of the lesson can be found in the *Power Maths* online subscription.

Reflect

WAYS OF WORKING Pair work

IN FOCUS This is a good opportunity to recap the 7 times-table. Some children may still rely on models such as number lines – these children will need interventions to progress towards mastery.

ASSESSMENT CHECKPOINT Assess children on their fast and fluent recall of the 7 times-table, when counting both on and back.

ANSWERS Answers for the **Reflect** part of the lesson can be found in the *Power Maths* online subscription.

After the lesson ⏸

- Could you set a home learning challenge?
- Would a class competition encourage children to learn their times-tables?
- Can children make links between the different times-tables?

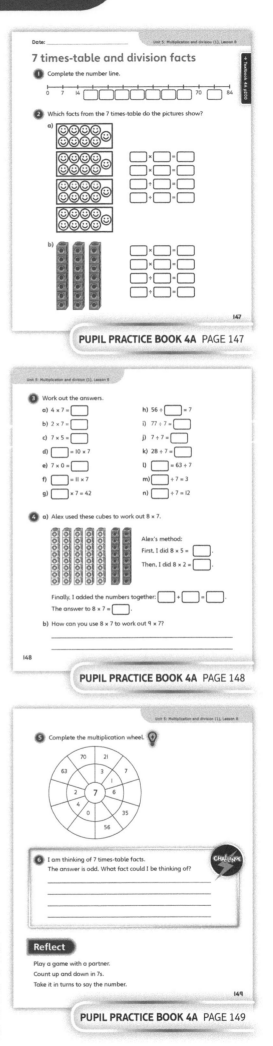

PUPIL PRACTICE BOOK 4A PAGE 147

PUPIL PRACTICE BOOK 4A PAGE 148

PUPIL PRACTICE BOOK 4A PAGE 149

11 and 12 times-tables and division facts

Learning focus

In this lesson, children will focus on learning their 11 and 12 times-tables. Children should be able to recall them quickly, and also learn the associated multiplication and division facts.

Before you teach

- How will you ensure children know all of their times-tables?
- Do children understand the links between different times-tables?
- How will you assess reasoning in this lesson?

NATIONAL CURRICULUM LINKS

Year 4 Number – multiplication and division

Recall multiplication and division facts for multiplication tables up to 12 × 12.

ASSESSING MASTERY

Children can demonstrate rapid recall of multiplication and associated division facts from the 11 and 12 times-tables. Children can use their knowledge to find solutions and clearly explain how they got the answers – this may include combining other times-tables, such as the 10 and 2 times-tables to reach the 12 times-table.

COMMON MISCONCEPTIONS

Children may learn the 11 times-table, but not understand why the number patterns occur, such as 11, 22, 33. Ask:
- *How could you use ten frames to represent the 11 times-table?*

Children may make mistakes when counting on or back in 11s or 12s. Ask:
- *How do you count on in 11s? How do you count back in 11s? How could you use a representation in cubes to help you?*

STRENGTHENING UNDERSTANDING

Children often struggle with the 12 times-table because of the larger numbers involved. Show them strategies to work out answers, the patterns involved and pair the facts with representations, such as arrays. Ask: *What patterns can you spot in the 12 times-table?* (12, 24, 36, 48: the second numbers are double the first.)

Multiplication and division facts must be constantly reinforced. Run interventions until children become fluent with their recall.

GOING DEEPER

Challenge children by linking the 11 and 12 times-tables. Ask children to find out how the 12 times-table can be worked out from the 11 times-table.

KEY LANGUAGE

In lesson: times-table, base 10 equipment, number facts, groups, number line, multiply (×), how many, multiplication wheel, divide (÷)

Other language to be used by teacher: number sentences, common multiple

STRUCTURES AND REPRESENTATIONS

Number lines, arrays, multiplication wheels

RESOURCES

Mandatory: cubes, base 10 equipment

Optional: counters

 In the eTextbook of this lesson, you will find interactive links to a selection of teaching tools.

Quick recap

As a class, count together in 10s from 0 to 100.

Discover

Unit 5: Multiplication and division (1), Lesson 9

WAYS OF WORKING Pair work

ASK

- Question ① a): *How do you know Max's number is in the 11 times-table?*
- Question ① a): *How do you know Alex's number is in the 12 times-table?*
- Question ① b): *Are there any numbers that appear in both the 11 and 12 times-tables?*

IN FOCUS Question ① a) is important because it focuses on finding facts from the 11 and 12 times-tables. In question ① b) children explore the 11 and 12 times-tables further by counting in multiples of 11 and 12.

PRACTICAL TIPS Show the 11 and 12 times-tables on your maths display. Cover up different parts each day. You could leave counters on a table for children to make arrays of the 11 or 12 times-tables.

ANSWERS

Question ① a): 3 × 11 = 33

3 × 12 = 36

Question ① b): 11 times-table: 0, 11, 22, 33, 44, 55, 66, 77, 88, 99, 110, 121, 132

12 times-table: 0, 12, 24, 36, 48, 60, 72, 84, 96, 108, 120, 132, 144

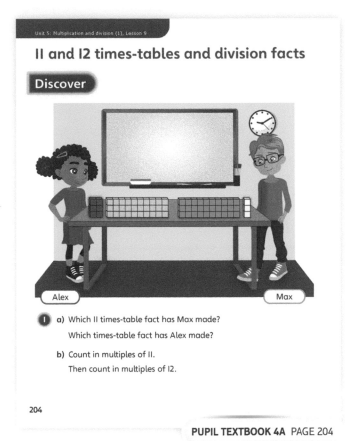

11 and 12 times-tables and division facts

Discover

① a) Which 11 times-table fact has Max made?
Which times-table fact has Alex made?

b) Count in multiples of 11.
Then count in multiples of 12.

204

PUPIL TEXTBOOK 4A PAGE 204

Share

WAYS OF WORKING Whole class teacher led

ASK

- Question ① a): *What is the difference between the patterns that Max made and the patterns that Alex made?*
- Question ① b): *Why does 132 appear in both groups?*

IN FOCUS For question ① a), ask children how they know Max's cubes represent a number in the 11 times-table. Ask them to create similar representations for numbers in the 12 times-table.

For question ① b), children can use the number line to count both on and back in multiples of 11 and 12.

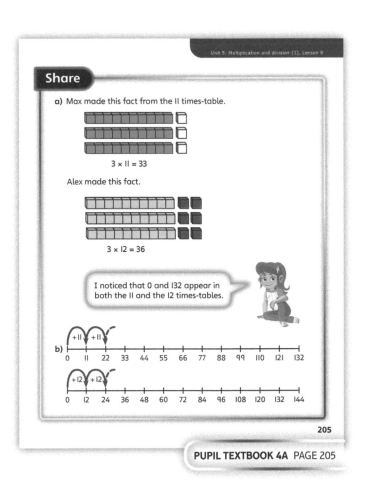

Share

a) Max made this fact from the 11 times-table.

3 × 11 = 33

Alex made this fact.

3 × 12 = 36

I noticed that 0 and 132 appear in both the 11 and the 12 times-tables.

b)

0 11 22 33 44 55 66 77 88 99 110 121 132

0 12 24 36 48 60 72 84 96 108 120 132 144

205

PUPIL TEXTBOOK 4A PAGE 205

Think together

WAYS OF WORKING Whole class teacher led (I do, We do, You do)

ASK

- Question ③: *Which multiplication and division facts do you remember?*
- Question ③: *Describe your method to a partner.*
- Question ③: *How will you check your partner's answers?*

IN FOCUS Question ③ is a superb activity for bringing together all of the prior learning in the unit. Children need to remember facts from a variety of times-tables.

STRENGTHEN For question ③, peer-assessing answers is a great way to strengthen learning.

DEEPEN Challenge children to create some word problems based on the 11 or 12 times-tables. Also, you could ask children to explain why 11×10 isn't 111.

ASSESSMENT CHECKPOINT Questions ① and ② will allow you to assess whether children can link multiplication and division facts to pictures or representations.

ANSWERS

Question ①: $7 \times 11 = 77$

Question ②: $4 \times 12 = 48$ $12 \times 4 = 48$

Question ③ a): Olivia multiplied $7 \times 10 = 70$ and $7 \times 2 = 14$ and then added the two together $70 + 14 = 84$. This is the same as $7 \times 12 = 84$.

Question ③ b): There is more than one way to solve these multiplications (but just one option below):
$4 \times 12 = 2 \times 12 + 2 \times 12 = 24 + 24 = 48$
$9 \times 12 = 3 \times 12 + 6 \times 12 = 36 + 72 = 108$
$60 \div 12 = 5$
$11 \times 12 = 10 \times 12 + 1 \times 12 = 120 + 12 = 132$
$84 \div 12 = 7$
$144 \div 12 = 12$

PUPIL TEXTBOOK 4A PAGE 206

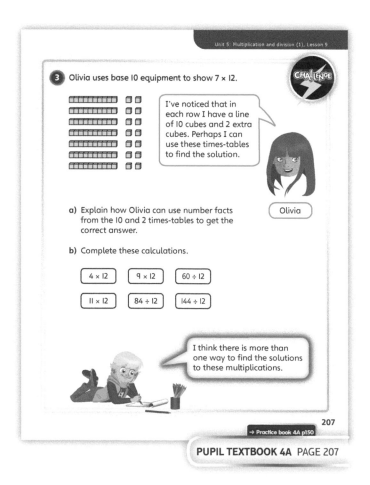

PUPIL TEXTBOOK 4A PAGE 207

Practice

WAYS OF WORKING Independent thinking

IN FOCUS Focus on question **5** with children. Finding related facts is an important activity which promotes the application of knowledge. This can lead to mastery of the topic.

STRENGTHEN Question **4** has number tracks in which some times-tables are in reverse. Some children may need to be reminded of this – revision of counting on and back might help here.

DEEPEN Challenge children to create word problems based on the 11 or 12 times-tables.

ASSESSMENT CHECKPOINT Run some spot-checks of all of the times-tables. Can children quickly recall them? Do not limit this to multiplication; include division facts too. Question **5** will give you a good indication of children's confidence, and mastery can be assessed based on whether children can find the related facts.

ANSWERS Answers for the **Practice** part of the lesson can be found in the *Power Maths* online subscription.

Reflect

WAYS OF WORKING Pair work

IN FOCUS Make this reflective exercise into a whole class competition. Children really enjoy times-table races – make it a weekly event – this will really encourage children to learn them.

ASSESSMENT CHECKPOINT Listen carefully to children's explanations of the strategies they used. Many will say: 'I just knew it'. Question this to encourage deeper thinking.

ANSWERS Answers for the **Reflect** part of the lesson can be found in the *Power Maths* online subscription.

After the lesson

- Did children complete the times-table grid quickly?
- Which times-tables did children find quickly?
- How will you continue to practise the times-tables with children?

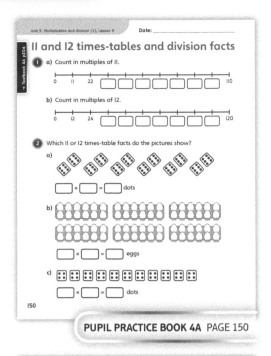

PUPIL PRACTICE BOOK 4A PAGE 150

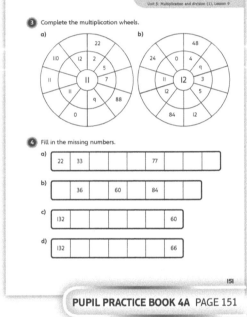

PUPIL PRACTICE BOOK 4A PAGE 151

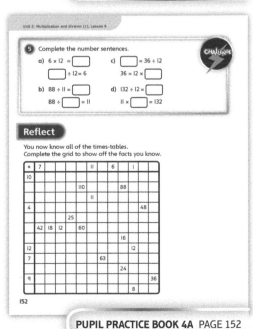

PUPIL PRACTICE BOOK 4A PAGE 152

Multiply by 1 and 0

Learning focus

In this lesson, children will learn how to multiply numbers by 0 and 1, finding out the rules and using visual representations to explain answers.

Before you teach

- How will you explain the rules for multiplying by 1 and 0?
- Do you need to recap arrays with children at the start of the lesson?

NATIONAL CURRICULUM LINKS

Year 4 Number – multiplication and division

Use place value, known and derived facts to multiply and divide mentally, including: multiplying by 0 and 1; dividing by 1; multiplying together three numbers.

ASSESSING MASTERY

Children can multiply a wide range of numbers by 1 and 0; they may use visual representations effectively to support their explanations. Children can explain their answers clearly.

COMMON MISCONCEPTIONS

Children may form the misconception that $327 \times 0 = 327$. Ask:
- *If you had 327 plates of 0 biscuits, would you have any?*

Children may add instead of multiply, for example, they state that $327 \times 1 = 328$. Ask:
- *Did you check the operation you are using?*

STRENGTHENING UNDERSTANDING

Children may need some extra practice multiplying by 1 and 0. You could run some intervention exercises in which children find solutions to real-life problems. Ask children to draw the calculations (or use counters) to strengthen their understanding.

GOING DEEPER

Challenge children by giving them calculations with more than one missing number. For example, $240 \times \square \times \square = 240$ or $240 \times \square \times \square = 0$. Ask them if they can find more than one answer.

KEY LANGUAGE

In lesson: multiply (×), zero, one, multiplication sentence, how many, groups of, grouped, in total, array, counters, function machine

Other language to be used by teacher: sets of, ones, (1s), tens (10s), hundreds (100s), lots of

STRUCTURES AND REPRESENTATIONS

Arrays

RESOURCES

Mandatory: counters

Optional: base 10 equipment

 In the eTextbook of this lesson, you will find interactive links to a selection of teaching tools.

Quick recap

As a class, count together from 10 to 0 on your fingers.

Discover

WAYS OF WORKING Pair work

ASK

- Question ❶ a): *How are the jam tarts grouped?*
- Question ❶ b): *Why do you think Emma is looking a bit upset?*

IN FOCUS For question ❶ b) children are required to think of the correct calculations in order to explain what mistakes Mo and Emma have made. It is important to tell children that they should use the calculations from question ❶ a) to help them.

PRACTICAL TIPS It will be helpful for children to be able to represent the plates full of tarts visually. Children could use toy plates and tarts or cakes to replicate the question. Real jam tarts on plates could also be used.

ANSWERS

Question ❶ a): There are 3 groups of 2. $3 \times 2 = 6$. Jamilla has 6 tarts.

There are 3 groups of 1. $3 \times 1 = 3$. Mo has 3 tarts.

There are 3 groups of 0. $3 \times 0 = 0$. Emma has 0 tarts.

Question ❶ b): Mo has added instead of multiplying. Mo says '$3 \times 1 = 4$'; but he should say '$3 \times 1 = 3$'. Emma has made a common mistake. She thinks multiplying by 0 is the same as multiplying by 1. Any number multiplied by 0 is always 0. Emma says '$3 \times 0 = 3$'; but she should say '$3 \times 0 = 0$'.

Share

WAYS OF WORKING Whole class teacher led

ASK

- Question ❶ a): *How do the arrays help you to understand the question and answer?*
- Question ❶ b): *What other mistakes could be made when multiplying by 1 or 0?*

IN FOCUS Question ❶ b) provides a good opportunity to support children with their mathematical reasoning skills. Suggest vocabulary that children can use in their explanations, and prompt them to give the correct solution.

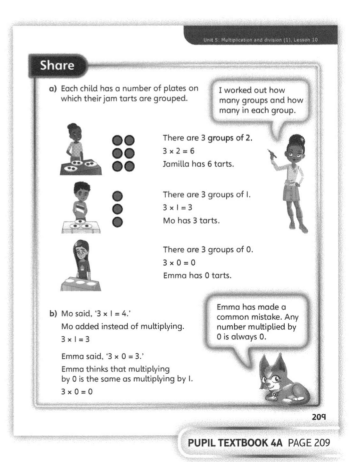

PUPIL TEXTBOOK 4A PAGE 208

PUPIL TEXTBOOK 4A PAGE 209

Think together

ASK

- Question ❷ c): *Are there any counters? Why not?*
- Question ❸ a): *Have you used the words 'change' and 'value' in your explanation?*

IN FOCUS In question ❷ b), children may realise that arrays can be arranged in two different ways: 3 × 1, 3 counters vertically or 3 counters horizontally. This is a good opportunity to talk about commutativity with children, for example 3 × 1 is the same as 1 × 3.

STRENGTHEN Children often make mistakes when multiplying by 1 and 0. Run intervention exercises in which children can use counters and baskets to represent a range of calculations and to reinforce the fact that multiplying by 0 is always 0.

DEEPEN To deepen learning in this section, ask children to think of some real-life problems in which the answer to a multiplication is 0. Ask: *A child has just finished 3 bags of sweets, how many are left in the bags?*

ASSESSMENT CHECKPOINT Question ❷ will allow you to assess which children can represent calculations visually.

Question ❸ is an excellent chance to assess children on their explanations of what happens when multiplying by 0 and 1.

ANSWERS

Question ❶ a): $5 × 1 = 5$

Question ❶ b): $5 × 0 = 0$

Question ❷ a): ●●● × ●●●● = ●●●● ●●●● ●●●● ●●●●

Question ❷ b): ●●● × ● = ●●●

Question ❷ c): ●●● × ____ = ____

Question ❷ d): ●● × ●●● = ●●● ●●●

Question ❸ a):
$5 × 1 = 5$	$6 = 6 × 1$	$10 = 10 × 1$
$1 × 15 = 15$	$17 × 1 = 17$	$1 × 183 = 183$

Mo notices that when you multiply a number by 1, it does not change in value.

Question ❸ b):
$5 × 0 = 0$	$6 × 0 = 0$	$0 = 10 × 0$
$0 × 15 = 0$	$17 × 0 = 0$	$0 × 183 = 0$

Emma notices that when you multiply a number by 0, the answer is always 0.

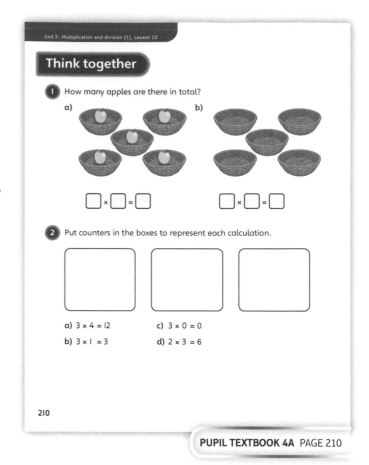

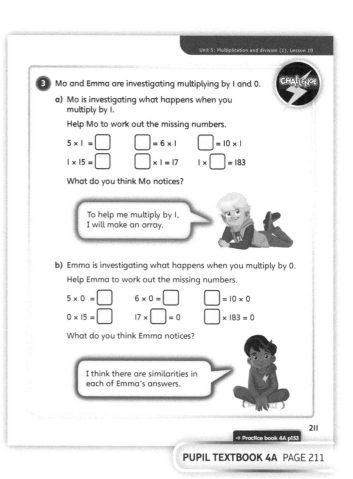

Practice

WAYS OF WORKING Independent thinking

IN FOCUS In question **2** c), children may put 0 × 0 = 0. It is important to explain to them that there are 4 trays with 0 cubes on, so the correct calculation is 4 × 0 = 0.

STRENGTHEN Sometimes children can be thrown by a larger number, like when calculating question **4** d). This is a good opportunity to show children that, even if they had 1 million multiplied by 0, the answer would still be 0.

DEEPEN Provide children with some more function machines with × 0 as one of the operations, like the one in question **5**, but only provide the answers. Ask: *What number went into the machine?*

After question **5** has been completed, discuss with children what would happen if the '× 0' appeared at the start or the middle of the machine.

ASSESSMENT CHECKPOINT Question **5** is an excellent activity to assess for mastery. Children who have fully understood the concepts of this lesson, will not even work out the full answers. Instead, they will see that the final operation of the function machine is 0, meaning every answer will be 0.

ANSWERS Answers for the **Practice** part of the lesson can be found in the *Power Maths* online subscription.

Reflect

WAYS OF WORKING Pair work

IN FOCUS If children are struggling with their explanations, ask them to draw calculations or use counters. You could also put some key phrases on the board: multiply, zero, stay the same, no change in value.

ASSESSMENT CHECKPOINT To assess for mastery, look for explanations, such as: *When you multiply any number by 0, the answer will always be 0; When you multiply a number by 1, the number will not change in value.*

ANSWERS Answers for the **Reflect** part of the lesson can be found in the *Power Maths* online subscription.

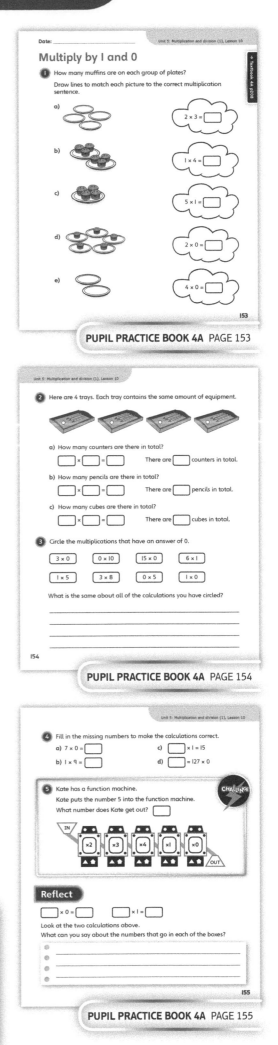

PUPIL PRACTICE BOOK 4A PAGE 153

PUPIL PRACTICE BOOK 4A PAGE 154

PUPIL PRACTICE BOOK 4A PAGE 155

After the lesson

- Would a quick recap of multiplying by 1 and 0 (at the start of the next lesson) cement understanding?
- Did all children understand that when you multiply any number by 1, it does not change in value?
- Did all children understand that any number multiplied by 0 is 0?

Divide by 1 and itself

Learning focus

In this lesson, children will learn how to divide numbers by 1. They will also relate their divisions to the inverse (multiplications).

Before you teach

- How will you make links between division and multiplication?
- How will you assess reasoning skills during the lesson?

NATIONAL CURRICULUM LINKS

Year 4 Number – multiplication and division

Use place value, known and derived facts to multiply and divide mentally, including: multiplying by 0 and 1; dividing by 1; multiplying together three numbers.

ASSESSING MASTERY

Children can confidently explain how to divide numbers by 1. Children can also link between dividing numbers by 1 and dividing a number by itself.

COMMON MISCONCEPTIONS

Children may subtract instead of divide, for example, 327 ÷ 1 = 326. Ask:
- *Did you check the operation you are using?*

Children may learn the pattern of dividing by 1, but not the understanding behind it. Ask:
- *Can you draw an array to help explain how to solve 12 ÷ 1?*

STRENGTHENING UNDERSTANDING

Children may need some extra practice dividing by 1. Run some intervention exercises in which children find solutions to real-life problems. For example, dividing 10 toy cows between one animal pen. They could draw arrays to support their conceptual understanding.

GOING DEEPER

Challenge children by giving them calculations which link multiplying and dividing by 1, such as $6 \div \boxed{} = 1 \times \boxed{}$.

KEY LANGUAGE

In lesson: divide (÷), calculation, how many, fact family, array, share, row, column, multiply (×), equal(ly), greater than

Other language to be used by teacher: lots of, groups of, sets of, ones (1s), tens (10s), hundreds (100s), total, less than

STRUCTURES AND REPRESENTATIONS

Arrays

RESOURCES

Mandatory: counters

Optional: base 10 equipment

 In the eTextbook of this lesson, you will find interactive links to a selection of teaching tools.

Quick recap 🔁

Ask children to show you how they would share 12 counters between 4, then between 3 and finally between 2. Ask: *What do you notice?*

Discover

Pair work

ASK

- Question ① b): *What is the difference between dividing a number by 1 and dividing a number by itself?*
- Question ① b): *How is multiplying by 1 similar to dividing by 1?*

IN FOCUS For question ① a), all of the questions are dividing numbers by 1, whereas question ① b) is dividing a number by itself. Ask children what the difference is between dividing a number by 1 and a number by itself.

PRACTICAL TIPS Link back to the previous lesson in which children had to multiply by 1. See if children can explain similarities between multiplying and dividing by 1. Make use of counters and 1 hoop by asking children to physically put them into 1 group.

ANSWERS

Question ① a): $5 \div 1 = 5$
$8 \div 1 = 8$
$3 \div 1 = 3$

Question ① b): $4 \div 4 = 1$

Unit 5: Multiplication and division (1), Lesson 11

Divide by 1 and itself

Discover

① a) How many bags of hay does Ted have?
How many bags of hay do Rose and Baz have?
What division could you use to work out how much hay each horse gets?

b) Write a division to share the hay between Jo's horses.

212

PUPIL TEXTBOOK 4A PAGE 212

Share

Whole class teacher led

ASK

- Question ① a): *What is similar about each of these questions and what is different?*
- Question ① b): *How many horses are there? How many bags of hay are there? What do you notice?*

IN FOCUS Question ① b) provides a good opportunity to explore the fact that when you divide a number by itself the answer is always 1.

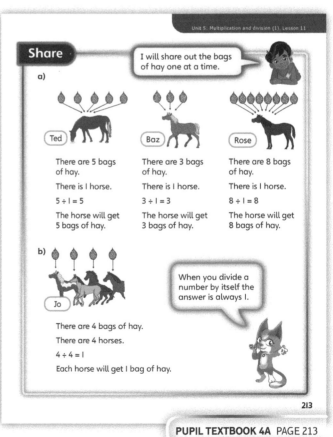

Share

I will share out the bags of hay one at a time.

a)

Ted There are 5 bags of hay.
There is 1 horse.
$5 \div 1 = 5$
The horse will get 5 bags of hay.

Baz There are 3 bags of hay.
There is 1 horse.
$3 \div 1 = 3$
The horse will get 3 bags of hay.

Rose There are 8 bags of hay.
There is 1 horse.
$8 \div 1 = 8$
The horse will get 8 bags of hay.

b)

Jo

When you divide a number by itself the answer is always 1.

There are 4 bags of hay.
There are 4 horses.
$4 \div 4 = 1$
Each horse will get 1 bag of hay.

213

PUPIL TEXTBOOK 4A PAGE 213

Think together

Whole class teacher led (I do, We do, You do)

ASK

- Question **1**: *Can you draw an array for this fact family?*
- Question **2**: *How does one fact help you to find the next fact in each of these fact families?*

IN FOCUS Question **3** asks children to reason why the divisions are linked. You may want to provide key words (divide, 1, itself) to support explanations.

STRENGTHEN Children may need extra practice with dividing by 1. Run some extra intervention sessions in which children have to sort counters into groups of 1.

DEEPEN Show children the difference between sharing and grouping by 1. Visual supports using counters or base 10 equipment and hoops will be helpful. For example, 4 ÷ 1 = 4: if you group it you would put 4 counters in 1 hoop, whereas if you share it out equally then you would have 4 hoops with 1 counter in each.

ASSESSMENT CHECKPOINT Question **2** will allow you to assess which children can link multiplication and division facts, when multiplying and dividing by 1.

Question **3** b) is a great opportunity to ask children if they understand why patterns occur when dividing numbers by 1 or itself.

ANSWERS

Question **1**: 10 × 1 = 10   10 ÷ 1 = 10
            1 × 10 = 10   10 ÷ 10 = 1

Question **2** a): 5 × 1 = 5, 1 × 5 = 5, 5 ÷ 1 = 5, 5 ÷ 5 = 1

Question **2** b): 3 × 1 = 3, 1 × 3 = 3, 3 ÷ 1 = 3, 3 ÷ 3 = 1

Question **3** a): 4 ÷ 1 = 4   4 ÷ 4 = 1
            5 ÷ 1 = 5   5 ÷ 5 = 1
            7 ÷ 1 = 7   7 ÷ 7 = 1
            10 ÷ 1 = 10   10 ÷ 10 = 1
            15 ÷ 1 = 15   15 ÷ 15 = 1
            32 ÷ 1 = 32   32 ÷ 32 = 1
           142 ÷ 1 = 142   142 ÷ 142 = 1

Question **3** b): When a number is divided by 1, its value stays the same. When a number is divided by itself, the answer is always 1.

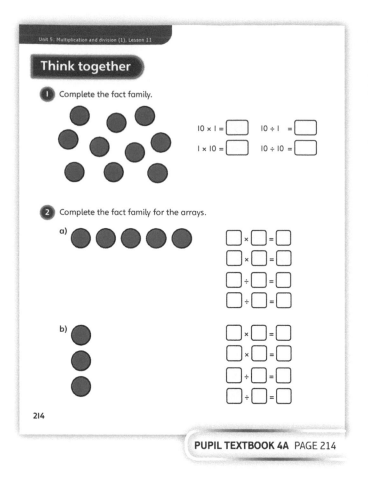

PUPIL TEXTBOOK 4A PAGE 214

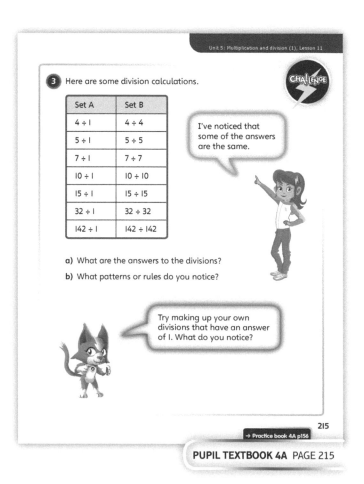

PUPIL TEXTBOOK 4A PAGE 215

Practice

WAYS OF WORKING Independent thinking

IN FOCUS For question ②, encourage children to explain their answer clearly, using an array if needed. This is also a good opportunity to explore what happens if you divide a number by 0.

STRENGTHEN Some children may not link dividing a number by 1 and dividing a number by itself. Support them by providing some extra intervention activities where they have to write fact families for an array.

DEEPEN Deepen learning by challenging children to complete number sentences with missing numbers, for example $7 \div \square = 1 \times \square$.

In question ⑥, challenge children to think of a range of numbers that could be represented by the shapes. Being able to complete this question correctly will show that children understand that dividing any number by 1 will give an answer of the same value because it does not need to be shared.

ASSESSMENT CHECKPOINT Question ⑤ is a great way to assess mastery of this lesson. Children will confidently complete the calculations and reason how they reached the answers when they are questioned.

ANSWERS Answers for the **Practice** part of the lesson can be found in the *Power Maths* online subscription.

Reflect

WAYS OF WORKING Pair work

IN FOCUS This section is a good opportunity to link back to the previous lesson's learning. Ask children what similarities they can see between dividing by 1 and dividing by the number itself.

ASSESSMENT CHECKPOINT To assess for mastery, look for children giving an explanation such as: *When you divide any number by itself, the answer will always be 1; When you divide a number by 1, the number will not change in value.*

ANSWERS Answers for the **Reflect** part of the lesson can be found in the *Power Maths* online subscription.

After the lesson ⏸

- Which children mastered the lesson?
- Will any children need a recap of the learning later on?
- Can children explain their calculations, backing up their explanation with arrays?

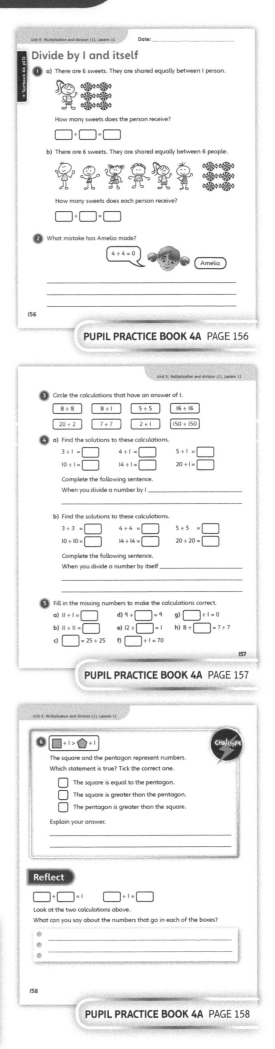

PUPIL PRACTICE BOOK 4A PAGE 156

PUPIL PRACTICE BOOK 4A PAGE 157

PUPIL PRACTICE BOOK 4A PAGE 158

Multiply three numbers

Learning focus

In this lesson, children will learn to find more efficient ways to multiply. They will use the commutative properties of multiplication to calculate 'in a different order', such as $2 \times 7 \times 5 = 7 \times 10$, to increase their ability to calculate mentally.

Before you teach

- Can children recall multiplication facts quickly?
- Do children know that multiplication is commutative?
- Do they know how to use arrays to multiply?

NATIONAL CURRICULUM LINKS

Year 4 Number – multiplication and division

Use place value, known and derived facts to multiply and divide mentally, including: multiplying by 0 and 1; dividing by 1; multiplying together three numbers.

ASSESSING MASTERY

Children use the properties of multiplication (that it is associative and distributive) and are learning to recognise the most efficient way to multiply three numbers. Children are able to recall multiplication facts rapidly.

COMMON MISCONCEPTIONS

Children may know the associative property of multiplication but fail to apply it to simplify when multiplying. For example, when multiplying $4 \times 7 \times 5$, they first multiply 4 and 7, and then use a written method to multiply the answer by 5. Ask:
- *How else could you work it out? Which numbers could you multiply first?*

Children may find it challenging to complete the second part of the multiplication, getting 'lost' when having three numbers to multiply. Increased fluency in multiplication facts will help. Ask:
- *What multiplication facts can you see in this multiplication? Are there any more?*

STRENGTHENING UNDERSTANDING

To strengthen understanding of multiplying three numbers, show the numbers using a visual or concrete representation, for example, two sets of arrays, both with 3 columns and 5 rows. Ask children to compare this with three sets of arrays, each with 5 rows and 2 columns. Ask:
- *What multiplication can you see? Which multiplication is easier? Why? What would happen if, instead of 3, you had a different number, such as 9? What would be easier to multiply: $5 \times 9 \times 2$ or $5 \times 2 \times 9$?*

GOING DEEPER

Ask children to reason why both sets of arrays described above show the same answer, without actually working out the multiplications. Ask: *If you joined the columns together so that there was no gap between the columns, what multiplication would you see?*

KEY LANGUAGE

In lesson: multiplication

Other language to be used by the teacher: multiplicand, recall, multiply, divide, product, equal, grouping, array, commutative

STRUCTURES AND REPRESENTATIONS

Number line, arrays

RESOURCES

Mandatory: cubes, counters

Optional: 4 times-table flashcards

 In the eTextbook of this lesson, you will find interactive links to a selection of teaching tools.

Quick recap

Ask children to choose a starting 1-digit number. Ask them to double it, and then to double it again. Discuss how this is a method for finding multiplication facts for the 4 times-table.

Discover

Pair work

ASK

- Question **1** a): *How many columns of stickers are there? How many stickers are there in each column?*
- Question **1** b): *How many sheets of stickers are there? What multiplication can you use to work out the number of stickers? Why?*

IN FOCUS Children are required to use their multiplicative reasoning in order to solve this problem. This is an opportunity to explore the commutative property of multiplying. To ensure children's concrete understanding of multiplying three numbers, it is important to link multiplying three numbers with their experience of using counters and arrays. Encourage children to rearrange the objects in different ways and to demonstrate why the answer does not change, regardless of the order of multiplying.

PRACTICAL TIPS Provide children with counters and ask them to practically solve the problem posed in the picture. Adapt the challenge by varying the numbers used in the question.

ANSWERS

Question **1** a): 2 × 5 = 10
There are 10 stickers on one sheet.

Question **1** b): 2 × 5 × 3 = 30, 2 × 3 × 5 = 30
There are 30 stickers, in total, on the teacher's desk.

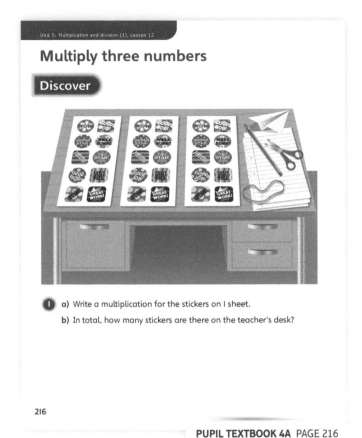

Multiply three numbers

Discover

1 a) Write a multiplication for the stickers on 1 sheet.

b) In total, how many stickers are there on the teacher's desk?

216

PUPIL TEXTBOOK 4A PAGE 216

Share

Whole class teacher led

ASK

- Question **1** a): *What does each group show?*
- Question **1** b): *Can you explain Dexter's method? Can you explain Flo's method? What do you notice about your answers? Which multiplication facts did you use? Why do you think it is important to know your times-tables off by heart?*

IN FOCUS Use this section as an opportunity to clarify any misconceptions that children may have. Encourage children to talk about the number of groups and the number of objects in each group. Question **1** b) provides an opportunity for children to explore the commutative property of multiplication and the variety of ways in which they can find the correct answer by multiplying three numbers in a different order. Ask children to use counters or arrays to show each multiplication and discuss how resources can be used in this way.

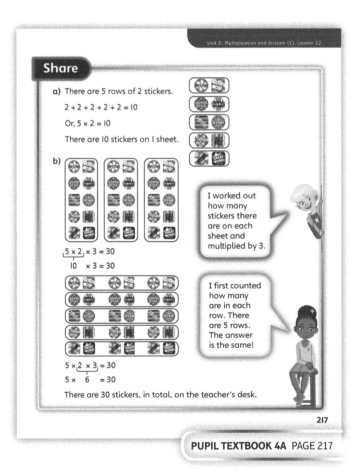

Share

a) There are 5 rows of 2 stickers.

2 + 2 + 2 + 2 + 2 = 10

Or, 5 × 2 = 10

There are 10 stickers on 1 sheet.

b)

5 × 2 × 3 = 30
10 × 3 = 30

I worked out how many stickers there are on each sheet and multiplied by 3.

I first counted how many are in each row. There are 5 rows. The answer is the same!

5 × 2 × 3 = 30
5 × 6 = 30

There are 30 stickers, in total, on the teacher's desk.

217

PUPIL TEXTBOOK 4A PAGE 217

Think together

Whole class teacher led (I do, We do, You do)

ASK

- Question **1**: *What facts can you see? How can you use these facts to work out the number of stickers? Can you multiply in a different way?*
- Question **2**: *How many doughnuts are there in each row? How many rows are there in each box? How did you multiply the numbers? Can you use a different method?*
- Question **3**: *Which multiplication facts do you find easier? Why? Look at Luis's and Isla's calculations. Which one is easier to do in your head? Which do you prefer?*

IN FOCUS In question **3**, children start to explore the commutative property even further and discuss the different methods they can use to multiply three numbers. They should notice that it is easier to multiply by 10, or to double a number, than to multiply by 3 or 9. Ask children to discuss what Astrid says and link the way children can add three numbers to the way they can multiply three numbers. Write '9 + 2 + 8' and '9 × 2 × 8'.

STRENGTHEN To strengthen understanding of multiplying three numbers, use visual or concrete representations, for example, with towers of cubes, or arrays. It is important that children can visualise and understand how to use their multiplication facts to multiply three numbers. Reinforce that they need to develop rapid recall of the multiplication and division facts, but also need to understand what they are. In question **2**, discuss Sparks's advice about drawing a diagram. Ask: *What diagram can you draw? How many groups will there be?*

DEEPEN To extend question **3** ask children to calculate 45 × 8. Encourage them to write 45 as '5 × 9'. Ask: *What is the easiest way to calculate this?* Write '5 × 9 × 8' on the board and ask: *Which numbers can you multiply first?* Provide other examples and encourage children to generalise that multiplication is commutative.

ASSESSMENT CHECKPOINT Children use their multiplication facts to multiply three numbers. They can work out which fact will help them to work out the answer in the easiest and quickest way.

ANSWERS

Question **1**: 5 × 2 × 6 = 60 or 2 × 5 × 6 = 60
     10 × 6 = 60
     There are 60 stickers on 6 sheets.

Question **2** a): 3 × 6 × 2 = 36; there are 36 doughnuts in 2 boxes.

Question **2** b): 3 × 6 × 5 = 90; there are 90 doughnuts in 5 boxes.

Question **3** a): Example answer: Isla's method is better as it is easier to multiply by 10 than by 5 or by 18.

Question **3** b): 7 × 6 = 42, 42 × 2 = 84
     4 × 5 = 20, 20 × 3 = 60
     9 × 8 = 72, 72 × 2 = 144

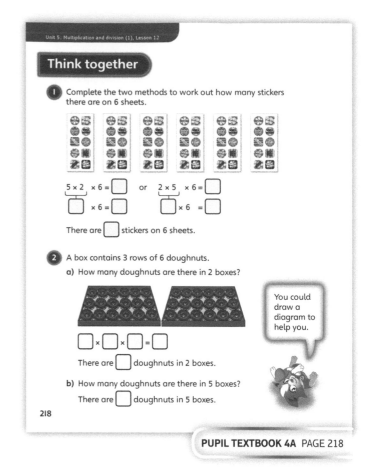

PUPIL TEXTBOOK 4A PAGE 218

PUPIL TEXTBOOK 4A PAGE 219

Practice

WAYS OF WORKING Independent thinking

IN FOCUS Question ❶ reinforces visual images of multiplying three numbers. It is important to cement this understanding. Question ❸ asks children to discuss the method of multiplying three numbers and to consider efficiency when choosing which numbers to multiply first.

STRENGTHEN To strengthen understanding of multiplying three numbers, show the numbers using visual or concrete representations. Question ❶ will help with this. As children move through the rest of the questions, they become more abstract, but children may still need to use cubes or counters to represent multiplications. It is important that children recognise the importance of knowing multiplication facts. Reinforce that they need to use their mathematical knowledge, not only to find a way to solve a question, but also to find the most efficient way to do so.

DEEPEN In question ❽, ask children to explain how they can find the missing numbers. They need to focus on what is given in the question and ask themselves how to use the information. Ask: *What is special about 60? Is it even or odd? What numbers can it be divided by? How do you know?*

THINK DIFFERENTLY In question ❼, children need to notice that the numbers are multiplied by 0, hence they do not need to multiply them, as the answer will be 0. Children need to explain their reasoning and mathematical thinking rather than guess the answer. This may be an opportunity to explain the misconception that a number does not change when multiplied by 0.

ASSESSMENT CHECKPOINT Can children find the most efficient way to multiply three numbers? They should have a secure recall of their multiplication facts (both multiplication and associated division facts) and use these to find the most efficient way to multiply three numbers.

ANSWERS Answers for the **Practice** part of the lesson can be found in the *Power Maths* online subscription.

Reflect

WAYS OF WORKING Independent thinking

IN FOCUS This brings together the commutative property of multiplying and knowledge of multiplication facts. Children should notice that it is easier to multiply by either 2 or 8 last: $(8 \times 5) \times 2$ or $(2 \times 5) \times 8$, rather than $(2 \times 8) \times 5$. Encourage children to take a strategic approach rather than just guessing. Ask them to explain their reasoning clearly.

ASSESSMENT CHECKPOINT Check whether children have instant recall of multiplication facts and are able to explain the properties of multiplication. They should be able to explain that multiplication can be done in any order because it is commutative. Check if there are any multiplication facts that they are confused about or any properties of multiplication that they do not understand.

ANSWERS Answers for the **Reflect** part of the lesson can be found in the *Power Maths* online subscription.

After the lesson ⏸

- Can children recall division and multiplication facts from the 12 times-table?
- Are they able to manipulate the facts they know in order to solve problems in an efficient way?

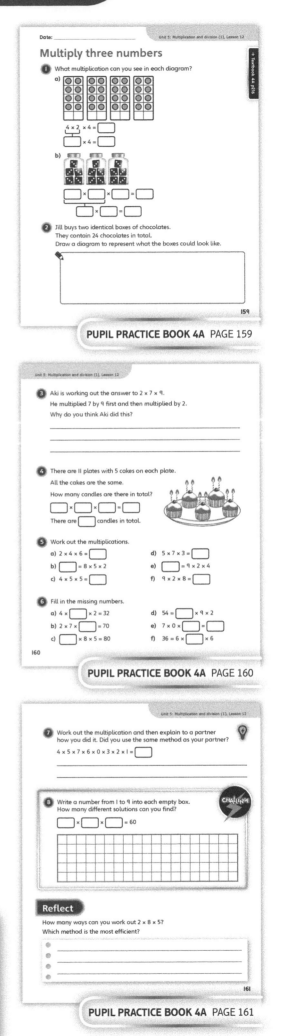

PUPIL PRACTICE BOOK 4A PAGE 159

PUPIL PRACTICE BOOK 4A PAGE 160

PUPIL PRACTICE BOOK 4A PAGE 161

End of unit check

> **Don't forget the unit assessment grid in your *Power Maths* online subscription.**

WAYS OF WORKING Group work adult led

IN FOCUS This end of unit check will allow you to focus on children's understanding of times-tables and whether they can apply their knowledge to find solutions.

- Questions **1** and **2** ask children to solve multiplications and divisions where numbers are missing.
- Look carefully at the answer that is given for question **3**. It will tell you if children understand how times-tables can be visually represented.
- Question **4** asks children to solve a multiplication involving the 12 times-table in a measurement context.
- Question **5** asks children to solve a multiplication of three numbers. Look for children using the commutative property of multiplication to find an efficient method.

Questions **6** and **7** are SATs-style questions, which will prepare children and get them used to the format.

ANSWERS AND COMMENTARY

Children who have mastered this unit will be able to quickly recall times-tables 1 to 12 (including multiplying by 0). They will know related multiplication and division facts and be confident matching multiplication and division facts to visual representations. Also, they will be able to find solutions to multi-step problems from their times-table knowledge.

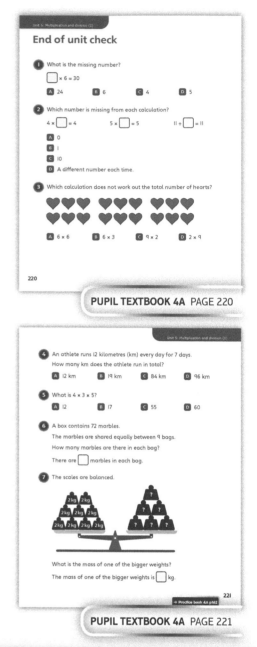

PUPIL TEXTBOOK 4A PAGE 220

PUPIL TEXTBOOK 4A PAGE 221

Q	A	WRONG ANSWERS AND MISCONCEPTIONS	STRENGTHENING UNDERSTANDING
1	D	A suggests children have done a subtraction rather than multiplication. B or C suggests children have made a times-table error.	Continual practice of multiplication and division facts will lead to mastery. Match multiplication and division facts with pictorial representations to strengthen knowledge of their meaning.
2	B	A suggests that children do not understand that multiplying by 0 always equals 0.	
3	A	C or D may suggest they do not realise that the hearts can be grouped in different ways, or that children do not understand the rules of commutativity.	
4	C	B suggests that children are adding instead of multiplying.	
5	D	A suggests children have only multiplied the first two numbers and B suggests they have added the third number. C suggests a multiplication error.	
6	8 marbles	Watch for children who have made errors when calculating – they may need their knowledge of times-tables strengthening.	
7	3 kg	Check children are able to see that both sides of the scales need to be equal.	

My journal

WAYS OF WORKING Independent thinking

ANSWERS AND COMMENTARY

Question **1**: Children will work individually to write their response, having discussed the problem in pairs or small groups beforehand. Encourage children to think through this section before writing their answer.

$45 \div £3 = 15$, so Jamilla could buy 15 small presents. Or, for example: 2 large presents $£9 \times 2 = £18$, 3 medium presents $£6 \times 3 = £18$ and 3 small presents $£3 \times 3 = £9$; $£18 + £18 + £9 = £45$. There are other correct answers. Children will need to use their knowledge of the 3, 6 and 9 times-tables.

Question **2**: Children may know the correct answer, but may find it challenging to write the reason why. Support children with key vocabulary to use and structure an answer for them.

A: $6 \times 7 = 42$

   7 books cost £42.

B: $48 \div 6 = 8$

   Each child receives 8 sweets.

C: $90 \div 9 = 10$

   I can buy 10 board games.

D: $2 \times 9 \times 9 = 162$

   9 bags weigh 162 kg.

Power check

WAYS OF WORKING Independent thinking

ASK

- *What times-tables do you know that you did not at the start of the unit?*
- *What kinds of multiplication and division problems can you do now that you could not at the start of the unit?*
- *What new words have you learnt and what do they mean?*
- *How can you use visual representations to show grouping and sharing?*

Power puzzle

WAYS OF WORKING Pair work

IN FOCUS Use this **Power puzzle** to assess children's knowledge and the speed of their times-table recall. Ask them to explain their methods or any strategies they used. Children take turns to time each other and then mark the answers.

ANSWERS AND COMMENTARY If children can do the puzzle successfully, then it means they can quickly recall multiplication and division facts. You will have to listen to the explanations of their strategies to assess whether they understand what they mean.

After the unit ⏸

- Have children got rapid recall of their multiplication facts?
- Do they know related division facts?

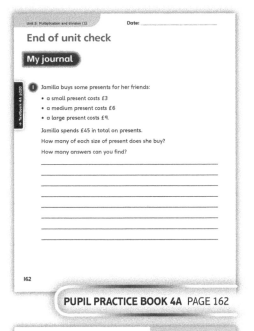

PUPIL PRACTICE BOOK 4A PAGE 162

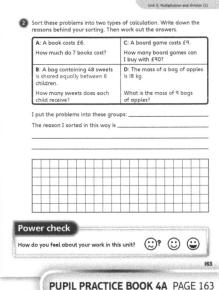

PUPIL PRACTICE BOOK 4A PAGE 163

PUPIL PRACTICE BOOK 4A PAGE 164

Strengthen and **Deepen** activities for this unit can be found in the *Power Maths* online subscription.

Published by Pearson Education Limited, 80 Strand, London, WC2R 0RL.

www.pearsonschools.co.uk

Text © Pearson Education Limited 2018, 2022
Edited by Pearson and Florence Production Ltd
First edition edited by Pearson, Little Grey Cells Publishing Services and Haremi Ltd
Designed and typeset by Pearson and Florence Production Ltd
First edition designed and typeset by Kamae Design
Original illustrations © Pearson Education Limited 2018, 2022
Illustrated by Laura Arias, John Batten, Paul Moran and Nadene Naude at Beehive Illustration,
Kamae Design and Florence Production Ltd
Cover design by Pearson Education Ltd
Back cover illustration © Diego Diaz and Nadene Naude at Beehive Illustration
Series editor: Tony Staneff; Lead author: Josh Lury
Authors (first edition): Tony Staneff, Josh Lury, Neil Jarrett, Stephen Monaghan, Beth Smith and
Paul Wrangle
Consultants (first edition): Professor Jian Liu and Professor Zhang Dan

The rights of Tony Staneff and Josh Lury to be identified as authors of this work have been
asserted by them in accordance with the Copyright, Designs and Patents Act 1988.

First published 2018
This edition first published 2022

26 25 24 23 22
10 9 8 7 6 5 4 3 2 1

British Library Cataloguing in Publication Data
A catalogue record for this book is available from the British Library

ISBN 978 1 292 45056 8

Printed in the UK by Ashford Press Ltd

For Power Maths online resources, go to:
www.activelearnprimary.co.uk

Note from the publisher
Pearson has robust editorial processes, including answer and fact checks, to ensure the
accuracy of the content in this publication, and every effort is made to ensure this publication
is free of errors. We are, however, only human, and occasionally errors do occur. Pearson is
not liable for any misunderstandings that arise as a result of errors in this publication, but it is
our priority to ensure that the content is accurate. If you spot an error, please do contact us at
resourcescorrections@pearson.com so we can make sure it is corrected.